PREFACE

In 1984, the Foundation commissioned a series of studies on the Legal, Technical and Safety Aspects relating to the Transport of Dangerous Wastes. These studies were undertaken at the request of the European Parliament and in support of the Commission's implementation of the Council Directorate (84/631/EEC) on the supervision and control within the European Community of the transfrontier shipments of hazardous wastes.

This initiative was followed up, in November 1985, by the organisation of a European Round Table on Safety Aspects of Hazardous Wastes, the purpose of which was to discuss the existing and future problems in this area, to bring forward new ideas and to help the Foundation in defining its activities 1985-88. A vast number of conclusions and recommendations emerged from this meeting and have since led to a series of studies on hazardous waste management being undertaken as part of the Foundation's work programmes 1986-88. Those on Safety Aspects relating to the Handling and Monitoring of Hazardous Wastes were among the first ones and were carried out in the twelve Member States in 1986-87. They focused notably on the existing legislation and regulations, as well as practices, aimed at both work-safety and environmental protection in this area, on internal safety procedures at establishments handling hazardous wastes and on the information required regarding the nature of the wastes, the potential risks and the measures in case of accidents. A number of case studies were undertaken as part of this work.

This report brings together the findings of the twelve national studies and analyses them in the wider European Community context. It examines the extent to which the existing legislation, regulations and practices are adequate, and it points to a number of shortcomings, both at the national and at the Community level. On this basis, the report makes a number of recommendations aimed at improving the efficiency and effectiveness of the present national legislation and at ensuring better practices and higher standards of waste management throughout the Community.

A meeting was held in Brussels, on 9 November 1989, to enable representatives of the employers, trade unions, governments and the Commission of the European Communities - the constituent bodies of the Foundation's Administrative Board - to evaluate the findings of this research. There was complete agreement on the value of the work that had been undertaken, and the present report was considered very useful, notably for the Commission. In relation to work safety, the participants felt that more emphasis should be placed on risk assessment and other preventive measures, and long-term health monitoring should be used to check the effectiveness of these measures. Moreover, training was seen as the essential prerequisite for obtaining higher standards of waste treatment and disposal, particularly the training of:

- the personnel of regulatory or enforcement agencies, to ensure, e.g., that the same standards are being used;

- managers of and operators at waste disposal and treatment facilities.

The participants also suggested a few minor amendments, and these have been included in the present report.

As a follow-up to this work, a series of studies on Education and Training relating to Hazardous Wastes was carried out in the twelve Member States in 1988-89. The findings of these studies are scheduled for publication in 1990.

Jørn Pedersen
Dublin, October 1989

Jørn Pedersen
Research Manager
European Foundation for the Improvement of
Living and Working Conditions

ACKNOWLEDGEMENTS

This report presents a synthesis of the key findings of twelve national studies on the Safety Aspects Relating to the Handling and Monitoring of Hazardous Wastes in the European Community. These studies were commissioned by the European Foundation for the Improvement of Living and Working Conditions and were carried out 1986-87.

ECOTEC Research and Consulting Ltd would like to acknowledge the valuable input into this project that has been made by the following individuals and organisations who prepared the national reports, which provided the information synthesised in this report.

R C Haines

Belgium	Mr Jean Pierre Hannequart Ms Marie Hannequart
Denmark	Mr Mogens Palmark Chemcontrol A/S
Federal Republic of Germany	Mr Ulrich H Kinner Dr Manfred Niclauss ECOSYSTEM GmbH
France	Mr René Goubier Agence Nationale pour la Récupération et l'Elimination des Déchets (ANRED)
Greece	Dr M Harris Mr E Pantzievits ECOTEC Research and Consulting Ltd
Ireland	Mr Matthew Lynch Institute for Industrial Research and Standards (IIRS)
Italy	Dr G U Fortunati
Netherlands	Ir H S Buijtenhek G J Kremers A Zonneveld C I Boeckhout TAUW - Infra Consult
Portugal	Mr José Antonio Martins Reis Techinvest SA
Spain	Mr Manuel de la Vega de la Rosa Mr Maximiliano Junquera Treisa SA
United Kingdom	Mr D Bardsley Dr M Harris ECOTEC Research and Consulting Ltd

CONTENTS

LIST OF TABLES

1.0 INTRODUCTION

1.1 Background and Rationale for the Study

In November 1985, the European Foundation for the Improvement of Living and Working Conditions held a European Round Table on Safety Aspects of Hazardous Wastes to help guide the Foundation in the definition of its research on this theme as part of its four-year rolling programme 1985-1988.

A number of the recommendations of the Round Table on Safety Aspects of Hazardous Wastes focused on various provisions and measures required to ensure the safe handling and monitoring of hazardous wastes in the European Community, particularly at waste disposal sites and waste treatment plants.

One of these recommendations related to various aspects of EC regulations concerned with the monitoring and control of hazardous wastes, and other relevant health, occupational hygiene (including safety at work) and environmental protection measures. The Round Table suggested that a comparative study be undertaken on the legislation and its enforcement in this area in the Member States of the EC and that it also include a review of the internal safety regulations at waste disposal sites and treatment plants. Another recommendation dealt with the requirement to provide, and assess the availability of, information on the nature of the wastes, the potential risks and the measures to be taken in case of accidents. It suggested that an assessment was made as to how it would be feasible to provide all persons involved in the handling of wastes with clear and relevant information in this respect, by means of a Community-wide waste notification system based on those systems used in a few Member States.

These recommendations were of particular interest to the Foundation as the studies they suggested are an important part of its integrated research programme in the field of hazardous wastes. They are also closely related to, and enlarged the scope of, other studies carried out by the Foundation. Examples include the Transportation of Toxic and Dangerous Wastes (carried out in 1984-1985), a 1986 project on contaminated land in the EC (undertaken in co-operation with the German Federal Ministry for Research and Technology and the Commission) as well as future studies on Hazardous Wastes and the Public, Education and Training relating to Hazardous Wastes and Planning Aspects relating to Hazardous Wastes. Moreover, the recommendations focused on an area where the Foundation has a particular expertise, i.e. the interface between living and working conditions.

In the light of the above, a study on Safety Aspects relating to the Handling and Monitoring of Hazardous Wastes was undertaken in all Member States so as to provide a detailed analysis of the situation in the Community, to assess the significance of hazardous waste in the Community and to indicate the common themes and issues which relate, as far as possible, to an overall EC view. A synthesis of the key findings of these studies is presented in this report.

1.2 Aims of the Studies

The aims of the studies undertaken in all Member States were :-

- to provide the Foundation, the Community institutions and the Member States with a comprehensive picture and a comparative analysis of the current situation in Member States in relation to legislation, regulations and other measures regarding safety issues associated with the handling and monitoring of hazardous wastes, as part of the Foundation's overall effort towards safe waste management policies and practices which take into account public concern in this area;

- to indicate possible alternatives to, and improvements of, existing measures and practices, e.g. by identifying procedures which have been implemented successfully in some Member States;

- to provide the Community institutions with information which may assist in their discussions on, and definition of, future waste management policies; and

- to obtain additional information on issues closely linked to previous and ongoing Foundation studies on hazardous waste and to establish a better basis for the formulation and implementation of future research on this topic.

National reports were prepared which followed closely a set of guidelines drawn up by the Foundation (see Annex III) and which set out the information to be obtained and analysis to be undertaken. The work was guided by a co-ordination group of experts and representatives of the Commission, trade unions, employers and the Foundation's Committee of Experts. Annex II provides a full list of those attending the co-ordination meetings.

The following paragraphs of this section of the report comment on the policy context of the work and place the problem of hazardous waste management in the EC and individual Member States in context, through a description of the legislation and arisings of waste.

Section 2.0 summarises the health and safety legislation applicable to the handling and monitoring of hazardous waste. Key aspects, including the scope of legislation, medical provision, worker participation/representation, training and enforcement are discussed. Conclusions and recommendations are presented.

Section 3.0 describes and reviews the national environmental legislation and regulations on wastes for each Member State, indicating where compliance with the EC Waste Directives is not achieved and/or where national legislation goes beyond the objectives and duties of these Directives. Problems of implementation and enforcement are discussed.

Case studies on the monitoring and handling of hazardous waste at both treatment and disposal facilities for each Member State are summarised in Section 4.0. Where applicable, comments are made relating to 'good' and 'bad' practice and the overall safety standards of national waste handling and treatment/practices are assessed.

The final section of this report brings the findings and conclusions of each section together, by discussing common themes and distinctions between the Community Member States. A final set of conclusions and recommendations are also presented.

1.3 Policy Context

There are three key elements to the policy context of this work at the level of the European Community. Firstly, the Third Environmental Programme and Action Plan expressed the Community's commitment to improve and maintain environmental quality. In this, it aimed to move beyond merely repairing environmental damage from, for example, the indiscriminate dumping of hazardous wastes, to a preventative approach in which, as far as possible, environmental objectives are integrated into those of other policies and actions so that they can be designed to exert minimal environmental and public health impacts.

Within this general policy approach, a number of environmental issues stand out as key problems on a Community scale, including the problem of pesticides and nitrates in soil and water, micro-pollutants, and problems of air pollution, in particular those caused by acid emissions. A long standing issue is the problem of dealing with waste in general, and hazardous waste in particular. There is insufficient information in the EC on the scale of arisings and trends and disposal practices utilised. In addition, there is evidence that there is a lack of technical knowledge in this area and the resources to cost-effectively handle and dispose of such wastes. This has led, in some cases, to poor standards of enforcement and monitoring.

Secondly, the Fourth Environmental Action Programme 1987-1992 reinforced the Community's commitment to a general preventative approach based on the achievement of high standards in all environmental sectors, through the development and utilisation of the necessary equipment, technologies, management and administrative policies (e.g. through the ACE regulation) and, at the same time, to find means of deriving economic and employment gains from such a move. With regard to Social Policy, the essential part played by environmental protection policy is recognised specially in the field of worker participation, professional education and general labour conditions. New actions are proposed specifically concerning the establishment and status of individuals responsible for industrial plant and the correct application of environmental protection regulations.

Thirdly, although there is no formal Community waste policy, principles and objectives have been set. These involve :-

- the harmonisation of systems of statistics on waste;
- clarification of the Community definition and nomenclature of dangerous waste:
- the development of a long term Community strategy on waste management; and
- the improvement of safety procedures covering movements of dangerous waste, with particular regard to professional training and the information given to haulage firms and drivers.

The work discussed in this report responds to the above issues by reviewing current practice in Member States, assessing aspects of handling and monitoring of hazardous wastes and the clean-up of contaminated land, and identifying key issues requiring further attention.

1.4 Hazardous Waste in Europe and its Specification

To set this work in its context, there is a need to define what is meant by hazardous wastes, to provide information on volume of arisings and trends and disposal practices. It is not the purpose of this work to provide a detailed review of the hazards and problems of hazardous waste as this has been well documented elsewhere.

There is no agreed definition within the Community of the terms "hazardous wastes", "toxic and dangerous waste" or "special wastes". Each Member State exercises control and collects statistics on the basis of different definitions, which again, are based on different systems. The studies carried out in Member States are based upon national definitions.

EC Legislation

The point of departure for this study is to assess the degree to which Member States have complied with Community founded legislation and to identify and assess areas where national legislation goes beyond the duties of the Directives.

In terms of Community legislation, the Framework Directive on waste, 75/442/EEC, places a general duty on Member States to take the necessary measures to ensure that waste is disposed of without endangering human health and without harming the environment. Though certain categories of waste materials are excluded from the scope of the Directive, 'waste' is defined as :-

> "any substance or object which the holder disposes of, or is required to dispose of pursuant to the provisions of the national law in force"

The four main mandatory elements of this Framework Directive are that :-

- competent authorities with responsibility for waste are to be appointed;
- waste disposal plans are to be prepared;
- permits from the competent authorities are to be obtained by bodies handling waste; and
- the 'polluter pays' principle is to apply.

The Framework Directive on waste was followed by the Directive on toxic and dangerous waste, 78/319/EEC. The main provision of this being that toxic and dangerous waste may be stored, treated and/or deposited only by authorised undertakings. The Directive also makes provision for plans to be made, records kept, transport controlled, inspections made and reports produced. It provides a definition of 'toxic and dangerous waste' as :-

> "any waste containing or contaminated by the substances or materials listed in the Annex to this Directive of such a nature, in such quantities or in such concentrations as to constitute a risk to health and the environment".

The Annex lists 27 substances and the Directive specifies wastes that are to be excluded such as radioactive waste, explosives and hospital waste.

The Directive places a general duty on Member States to ensure that toxic and dangerous waste is disposed of without harming human health or the environment and, in particular, without risk to water, air, soil, plants or animals. A general duty also encourages waste reduction and re-use of toxic waste.

The notification date of this Directive was the 22nd March 1978, with normal compliance due on the 22nd March 1980. For some Member States this has caused few difficulties as national legislation covering general aspects of toxic and dangerous wastes, and the framework Directive were already in place. However, seven years after the deadline for compliance, some Member States have still not fully implemented the Directives and infringement procedures are being carried out. A detailed discussion of the legislation and compliance is contained in section 3.0.

Definitions of Hazardous Waste

Some difficulties in compliance are largely due to the imprecision of the definition of toxic and dangerous waste contained in the Directive, which has lead to wide interpretation and has resulted in a variety of working definitions within the EC. In some Member States, the issue of compliance with the Directive itself is in doubt because of the ambiguities and interpretations of the national toxic and dangerous waste definitions. For example, the UK defines toxic and dangerous wastes (special wastes) as those wastes dangerous to human health, omitting from the definition references to wastes that may be damaging to the environment. However, the Directive clearly states that the definition should include substances that constitute a risk to the environment. Some Member States have difficulty where strong regional administrations (e.g. in Belgium and Germany) have independently drawn up their own definitions of toxic and hazardous waste.

Whereas in some Member States full compliance with the letter and spirit of the Directive's definition is in doubt, others implement a toxic and dangerous waste legislation definition far more comprehensive than that of the Directive. For example, in Denmark 51 types of hazardous waste are defined, and a classification system, based on industry source is also operated. In Italy concentrations of certain hazardous substances, a list of hazardous waste producing activities and a list of defined substances are all parameters used to define hazardous wastes. Various other definitions are operated within the EC. A comparative summary of each of the national definitions is given in Table 1.1.

The position within the EC 12 with respect to the toxic waste definitions can be summarised by grouping together those Member States which operate similar definitions. It should be noted that the terminology used by Member States varies and terms such as chemical waste, special waste, special industrial waste and hazardous waste are, in the context of the report, equivalent to 'toxic and dangerous waste' as specified in Directive 78/319/EEC.

At the simplest level, Portugal and Greece use the toxic and dangerous waste definition of Directive 78/319/EEC. They have not expanded the list of dangerous substances contained in Annex I of the Directive, or introduced any concentration criteria in order to classify waste as toxic or non-toxic.

TABLE 1.1 : VARIABILITY OF HAZARDOUS WASTE DEFINITIONS USED BY MEMBER STATES

Country	Extended List of Annex 1 78/319/EEC	Defines Physical & Chemical Criteria for Hazardous Waste	Defines the Concentration Criteria of Substances	Defines Haz. Waste Production Processes	Regional Variations in Definition or Interpretation
Belgium	*	*	*	*	■
Denmark	■	■	*	■	*
France	■	*	*	■	*
Germany	*	*	*	■	■
Greece	*	*	*	*	*
Ireland	■	*	*	■	■
Italy	■	*	■	■	■
Luxembourg	■	*	■	*	■
Netherlands	■	*	■	■	*
Portugal	*	*	*	*	*
Spain	■	■	■	(■)	■
UK	*	■	*	*	*

■ Indicates adoption of this aspect into the national definitions

(■) Indicates adoption of this aspect into future regulations

The definitions used in the remaining 10 Member States vary significantly. Ireland has no hazardous waste definition formally incorporated into national law but interprets the Directive's definition in its widest sense. Other Member States extensively use the physical and chemical properties of waste to legally classify a waste as dangerous or hazardous. Others (e.g. Denmark) have extended the substances list, or have specified concentration criteria and hazardous waste producing processes.

Though the legal definitions used by Member States may affect the operational practice of sampling, monitoring and disposal of such wastes, the stringency or precision of definition does not necessarily ensure good control in hazardous waste management or that all hazardous wastes are covered by the definition and thus by the requirements of the legislation. In Italy, for example, substances, waste producing industrial processes and concentration criteria are all defined. Such a system calls for a considerable degree of analytical testing, which results in delays and higher treatment and disposal costs, all of which may be counter productive to the extent that waste producers may be encouraged to avoid formal procedures during the disposal of their wastes.

The lack of a national definition in Ireland has not, however, inhibited control because there are several laws relating to the disposal of waste and a non-legal memorandum, issued by the Irish Department of the Environment, has been generally adopted and implemented by industry as a quasi-legal document. The memorandum includes an extensive list of types of waste, to which local authorities refer when determining if a waste is toxic and dangerous, and it is not necessarily required that one or more of the 27 substances of the Directive be present. In addition, in practice the Memorandum list of wastes is not considered to be exhaustive, i.e. local authorities can designate other wastes as toxic or dangerous that are not contained on the list.

Thus the lack of rigid legislative control and definition can allow control authorities to adopt a wide interpretation, in some cases tighter than that given by Directive 78/319/EEC.

Some definitions that appear precise and without ambiguity can, when put into practice, be narrow in scope and limited in effect. For example, the specification of a concentration value (g/kg) may have no regard to the effect of the total quantity of waste being disposed of. Thus, although a substance concentration may be below the threshhold, and under the terms of the definition non-hazardous, the total waste load may contain a substantial quantity of the contaminating material capable of causing considerable environmental damage.

Proposal for a Council Directive on Hazardous Waste

During the course of the research for this study, the Commission of the European Communities has been discussing proposals for amendments to the above. Experience gained at Community level in waste management and, more importantly, the management and disposal of hazardous wastes has demonstrated that some provision of the Directive 75/442/EEC[*1] of 15th July 1975 on waste needs to be amended and Directive 78/319/EEC[*2] of 20th March 1978 on toxic and dangerous waste needs to be replaced. In addition, amendments of form are required to Directive 84/631/EEC[*3] of 6th December 1984 on the supervision and control of the transfrontier shipment of hazardous waste. Of particular relevance to this study are priorities to harmonise waste disposal methods and to introduce more precise and more consistent definitions of "waste", "disposal" and "hazardous waste" at Community level.

As discussed above, the definitions contained in the Directives have not, in some cases, been incorporated in national legislation and has led to administrative and technical problems over the safe treatment and disposal of wastes. Moreover, considerable delays have occurred in the implementation of Directive 84/631/EEC due to the lack of any precise and uniform community definition of hazardous waste.

The new proposals are for a Directive on hazardous waste, this term being wider and more comprehensive than the former "toxic and dangerous" waste. The definition of "hazardous" waste will be qualified by reference to three Annexes: a list of types of categories of hazardous waste, a list of substances or materials which render a waste hazardous and a list of hazard characteristics. This proposed definition is very close to the one formulated by the OECD Waste Management Policy Group.

*1 OJ < 194, 25.7.75, p39

*2 OJ < 34, 31.3.78, p43

*3 OJ < 326, 6.12.84, p31

1.5 Arisings of Hazardous Waste in the Community

The bulk of toxic and dangerous wastes are generated by the process industries, the main producers being the chemicals sector and the mineral and metal processing industries. Other industrial sectors, laboratories and hospitals also produce significant quantities. These wastes can arise as unwanted by-products through cleansing processes and from waste treatment or recovery processes, or due to accidents and spillages. They can occur as relatively pure substances or as complex mixtures in small and large quantities and as gases, liquids, slurries or solids.

Data on Arisings

Accurate time series data on national arisings and disposal routes of hazardous waste would be useful in identifying :-

- Whether hazardous waste is an increasing or decreasing problem

- Those countries where arisings are increasing and, therefore, where treatment and/or disposal capacity may need to be expanded so that such wastes can be handled in a safe and economic way

- Regions and countries which are being used as principal receptors of waste

- The main hazardous waste producing industries within these countries so as to identify R & D priorities to enable cleaner technologies to be developed and thus reduce the quantities of hazardous waste arising for disposal

- Those industries where arisings have declined either through recycling, cleaner processes or alternative materials.

From this, comparisons could be made between industries in different sectors and between Member States, to determine if they are making similar efforts and attaining the same objectives with respect to recycling, prevention and processing of waste. Arising figures would also enable the structure of transfrontier movement and disposal of hazardous waste to be more clearly defined.

The advantages of having such data are thus numerous, and important for the determination of future national and EC policy objectives on hazardous waste management. However, though data are available, indeed, the legislation requires records to be kept on the quantity, nature, physical and chemical characteristics and origin of waste by the competent authorities of Member States, the data are not of uniform quality throughout the EC.

Several factors affect the accuracy of the hazardous waste arising data and thereby reduce the utility of the information and the interpretation that can be made from it. These include :-

- The wide variation in national definition for hazardous waste used by member states for data collection purposes

- Reluctance by industry to supply data

- Poor data collection methods, (often based on rough estimates) and infrequency of data collection and hazardous waste arising surveys

- The degree to which 'hazardous waste' under the terms of individual national definitions is 'hidden' by industry and not recognised by competent authorities.

These aspects are discussed further in order to indicate the type of problems associated with obtaining accurate and comparable data on hazardous waste arisings.

Volume of Waste Arisings

As discussed above, the scope of definitions varies significantly in Member States and this influences the data available on hazardous waste arisings. For example, in Portugal where the definition of Directive 78/319/EEC has been legally adopted, a recent inventory of toxic and dangerous waste has been carried out. However, the inventory used a much broader concept of the definition than that provided by the Directive. Thus a discrepancy could exist between the amounts of hazardous waste subject to legal control compared to the quantity of waste apparently produced by the country according to survey results. Based on the wider concept, it is estimated that 1 million tonnes of toxic waste is generated every year by various industrial processes in Portugal. By 1995 the figure is forecast to rise to 2.2 million tonnes, not only because of the growth of manufacturing sectors but also as a result of new restrictions on uncontrolled discharges of industrial liquid effluent. At present, the bulk of hazardous wastes produced in Portugal is in the chemical sector which is responsible for 80% of all hazardous wastes generated.

It is estimated that 2 million tonnes per annum of hazardous waste are produced in France (out of a total of 18 million tonnes of special waste). However, the wider definition of 'special wastes' encompasses any waste containing a toxic substance and, therefore, this is perhaps a more useful indication of the scale of the hazardous waste management problem in France. Similar circumstances occur in other countries. In the UK, 'special waste' arisings (which are equatable with 'Toxic and Dangerous Waste') are approximately 1.5 million tonnes per annum. However, it is estimated that a total of 3.5 million tonnes of 'hazardous waste' arose in England alone for disposal in 1985.

In the Netherlands, toxic and dangerous wastes as defined by the Chemical Waste Act are specified by the concentration of certain toxic substances and/or by the industrial process from which they eminate. Thus, it is a more precise definition than that contained in the Directive. Under this definition, it is estimated that 1 million tonnes of chemical waste are generated each year in the Netherlands.

These examples illustrate the difficulty of trying to calculate the volume of arisings within the EC when the 'quantity' being measured is not uniformly or precisely defined. Nevertheless, an estimate of arisings has been made based upon the national reports and a summary of hazardous arisings is given in Table 1.2, together with the principal disposal routes. It should be noted that different definitions have been used and thus caution is required when making comparisons. The volume of arisings of hazardous waste in the EC in recent years is estimated to be 16.8 million tonnes per annum. However, it should be noted that this figure is based on national arisings for different years (see Table 1.2). To put the arising of hazardous waste into context, it is currently estimated that the total arisings of all wastes in the EC is some 2,200 million tonnes a year.

TABLE 1.2 : HAZARDOUS WASTE ARISINGS - PRINCIPAL SOURCES AND DISPOSAL ROUTES

Country	Quantity of Toxic and Dangerous Wastes		Principal Sources or Types of Waste	Principal Disposal Routes
	'000s tonnes	Year		
Belgium	1000	1987	Cleaning wastes, sludges, mixed organics and inorganics	Landfill treatment, and incineration
Denmark	128	1986	Oil wastes	Physical, chemical treatment and incineration
France	2000	1987	Organic chemical and mineral wastes	Landfill, incineration and treatment
Germany	4800	1983	Sulphur containing waste, acids, solvents and sludges	Landfill (31%), dumping at sea (18%), incineration (8%), treatment (7%)
Greece	*	*	Sludges and liquid effluent	Landfill (90%) (new specialist treatment/disposal capacity is planned
Luxembourg		1986	Organic wastes, flammable substances, and acids	Landfill
Ireland	58	1984	Organic and chlorinated plants, lead compounds, asbestos and acids	Export (for treatment and disposal - 28%), on-site treatment (66%), including recycling and incineration
Italy	710	1984	Manufacturing industry	Landfill and incineration
Netherlands	1000	1985	Chemical, petrochemicals and metals industry	Treated (in-house) (55%), incineration (10%), treated/ recycled (20%), controlled landfill (5%), export (10%)
Portugal	1050	1987	Tanneries, pulp and paper, iron and steel industries	Uncontrolled landfill (new specialist treatment/disposal capacity is planned)*
Spain	1800	1988	Chemical, pulp and paper and metal processing industries	Uncontrolled landfill (93%) incineration, treatment recycling (7%) (new specialist treatment/disposal capacity is planned)
UK	1550	1985	Manufacturing industry	landfill (79%)
Approx Total	**16870**			

* *No data available*

Note that the definition of controlled landfill is not consistent throughout Member States and what is regarded as controlled in one State may not necessarily be acceptable in another. Therefore, where landfill is indicated in the above table, the reader should not make any qualitative conclusions or comparison regarding the standards of landfill disposal within each Member State.

The estimated annual breakdown is as follows :-

120 million tonnes of household waste
950 million tonnes of agricultural waste
160 million tonnes of industrial waste
300 million tonnes of sewage sludge
250 million tonnes of waste from all extractive industries
170 million tonnes of demolition waste and debris
12 million tonnes of consumer waste (scrap vehicles, tyres etc)
200 million tonnes of other waste (foliage, litter, etc)

Thus, arisings of hazardous wastes account for approximately 0.7% of all EC waste and 8.8% of industrial waste. However, if a wider interpretation of the toxic and dangerous waste definition, which has been adopted by some countries, were used throughout the EC, then it is estimated that the hazardous waste arisings could be as high as 70-80 million tonnes per annum or 50% of industrial wastes arising.

An annual arising of some 70-80 million tonnes of hazardous waste (assuming the adoption of a wider definition) may not directly lead to any changes in the disposal practices of Member States, but it will substantially increase the administration workload of the industry and the competent authorities. In particular, the requirements for waste monitoring will increase if authorities are to maintain satisfactory records on all hazardous waste arisings and movements and ensure safe disposal is carried out.

In addition, existing disposal facilities authorised to receive hazardous waste may have insufficient capacity to accept the much higher quantities of hazardous waste. Thus waste acceptance criteria at many disposal sites may need to be re-defined in order that they can accept certain categories of 'less' hazardous waste. This change will require an increase in the level of waste sampling prior to disposal if wastes are to be disposed by the most appropriate method. It is also likely to be met with concern from environmental pressure groups because of the perceived shortage of adequate disposal facilities for such large volumes of waste.

Therefore, it is envisaged that major difficulties may arise if a wider interpretation of the toxic and dangerous waste definition were to be applied throughout the EC. In fact, rather than public and environmental safety being enhanced it is possible that the opportunities for inappropriate disposal by less reputable contractors may increase as the waste disposal authorities find that they are unable to satisfactorily monitor waste movements and disposal. This outcome seems even more likely, particularly as existing standards of hazardous waste management are far from satisfactory.

Trends in Arisings

Trends in hazardous waste arisings are difficult to assess, not only because of the lack of data but also because of shifting definitions. However, data for some Member States suggest that the quantity of specific types of waste are decreasing, either due to changes in production processes or through material substitution. Examples of this would be PCBs, where production has largely stopped, and solvents where recycling and replacement is taking place. In contrast to this, other wastes appear to be increasing, e.g. waste oils, pigments, and organic synthetic wastes. It is also recognised that the decline of certain polluting manufacturing sectors and the growing cost of waste disposal have also had some influence on the nature and quantities of hazardous waste arisings in the short term. Thus, it is the arisings of specific hazardous wastes which can be expected to change rather than the overall volumes in the medium term.

Disposal Routes

The principal disposal routes for hazardous wastes are also shown in Table 2 and include landfill, physical and chemical treatment, incineration and indefinite storage and. From Table 2 it is apparent that there is considerable variation in disposal practices within the EC. For example, in Denmark, the majority of wastes are either treated or incinerated at one central plant, KOMMUNEKEMI, whereas in the UK, 79% of the total disposals of hazardous waste are by controlled landfill (co-disposal). In other Member States, a variety of disposal routes are used. For example, in France approximately 20% goes to landfill, the rest being either incinerated, treated or recycled. This variation in disposal routes between the Member States arises for several reasons. These include :-

- National policies on waste management
- The availability of sites and facilities for disposal
- Hydrogeological and other physical factors
- The influence of industry and costs of disposal
- The influence of environmental protection lobby groups
- The quantity of arisings.

The importance of these individual factors in determining disposal routes similarly vary between Member States. In the UK, for example, there are a large number of sites that are regarded as suitable for the co-disposal of hazardous waste. This method is deemed to be acceptable by Central Government and because it is relatively cheap, industry also favours this route. In contrast, the Netherlands has few such sites and concern about the contamination of groundwater supplies in certain areas means that disposal by landfill is prohibited in general. In Denmark, the quantity of hazardous waste arisings by EC standards is small and a management system based on one central disposal/treatment facility has been established, which accommodates all hazardous waste arisings.

Although waste disposal routes show considerable national variations, in the context of the EC, landfilling remains the predominant disposal route with over 45% being disposed by co-disposal, controlled or uncontrolled tipping methods. Incineration and physico-chemical treatment are also significant disposal routes. However, it should be noted that a large proportion of the residues of treatment (contaminated ash, dewatered chemical sludges etc) are eventually disposed to landfill. The information available suggests that in the future, given tighter regulatory control over hazardous waste disposal, landfill will decline in significance as other, more environmentally appropriate, methods are utilised (e.g. incineration and chemical treatment).

Data Collection

Data collection procedures and the frequency of collection show considerable variation in the EC 12. In the UK, there has been no systematic survey of waste arisings and estimates that have been produced have been based upon local knowledge and from waste notification procedures. Similarly in Greece, there is no operational, centralised procedure for the collection of data on the quantity and nature of waste arisings.

In Spain, the most recent comprehensive figures for arisings were compiled in 1981. The Autonomous Communities in Spain have drawn up their own inventories of waste which are not considered to be reliable, either because of the financial constraints facing the bodies responsible for them, or because industries are unwilling to supply the required information. In 1981, it was estimated that 1.5 million tonnes of what were then defined as special wastes were produced. This figure is not considered to be an accurate guideline to current waste arisings, not only because it is seven years old, but also because this original estimate was based upon estimates provided by industry. Some Autonomous Communities have toxic waste arisings data for 1985. Again these data are not considered to be very accurate. For those Communities with 1985 data, however, all these figures are substantially higher than for 1981. The most recent estimation of arisings (*Dirección General del Medio Ambiento*) of toxic and dangerous waste for 1988 is 1.8 million tonnes.

A problem which affects the data quality in some countries is that where records are kept the units of measurement are not comparable or convertable. For example, in the UK and Ireland waste loads may be recorded in numbers of tonnes, litres, skipfuls, lorry loads, drums or tanker loads. Of these units, few have established magnitudes and significant errors can occur in the summation of the waste arisings expressed in these terms.

The data on toxic and dangerous arisings may be obtained by several methods. These include notifications provided by waste producing authorities, from questionnaires completed by industry and from records kept by waste producing companies. The utility of the data is determined by the coverage of the data collection exercise, the accuracy of the information supplied and the questionnaire sample size. The accuracy of the information supplied is affected not only by the measurement problems already mentioned, but also because the waste producers may not categorise their waste uniformly or declare all the wastes they produce. The resources allocated to waste inventories, i.e. handling and personnel involved, also influence the data quality of waste inventories.

Some countries have systems which facilitate the collection and compilation of waste arisings. In France, the intention is to have computerised monitoring of waste arisings and a statistical survey which will record the production, transportation, disposal and importing of wastes is currently being set up. In Denmark, industries producing chemical wastes are required to notify the relevant municipality of the quantity of arisings. This information is continously updated and thus national wastes arising are calculated from these decentralised sources.

Differences between notified waste quantities and the amount delivered to Kommunekemi have been observed in some cases. A computerised data bank system is therefore currently being established in Denmark. The system contains data on potential waste generators and information on generators who deliver hazardous waste to Kommunekemi. The database will enable the identification of discrepencies between waste loads which have been notified for disposal, but for which there is no treatment or disposal record. The system also makes it possible to contact generators who, at present, do not notify and/or deliver wastes to Kommunekemi.

Though the toxic waste Directive 78/319/EEC includes the provision that records be kept of the quantity, nature, physical and chemical characteristics and origin of waste, there is evidence that this system of record keeping, which should enable national statistics on waste arisings to be obtained and aggregated, does not currently ensure such figures are available. The UK operates the transport notification system, but there is no system or agency involved in compiling the records maintained by the Waste Disposal Authorities, and many of the Authorities themselves do not keep an up-to-date record. Thus, in the majority of Member States, inventories are the only means of assessing the total quantities of waste produced and, in general, these are not carried out on a systematic basis.

Transportation of Hazardous Waste in the EC

The transportation of hazardous wastes within the EC is principally by road. Ship, barge and rail transport are also used but, in terms of quantities transported, they are not significant. The dominance of road transportation arises because of the structure of the waste disposal industry. For example, the location of landfill sites is determined by past mineral winning industries to provide 'holes in the ground', and hydrogeological and geological factors and, thus, it is unlikely that such sites will be found close to a major canal, river system or a rail network. Road transport is often the only method of reaching landfill sites. Thus, wastes transported by rail or barge requires double handling before it reaches its final treatment or disposal point. Because the quantities of toxic and dangerous wastes produced by individual industries are usually quite small in comparison to the quantities of non-hazardous waste, their transportation is usually only practicable and economic by road.

The distances travelled by vehicles carrying toxic and dangerous wastes vary from country to country. Luxembourg has no specialist treatment facilities so all wastes have to be exported to neighbouring states. Similarly, Ireland also has a shortfall in treatment capacity and must export large quantities. To an extent, the more specialised the treatment method required for a particular waste the greater the distance the waste will have to be transported. For example, in the UK, there are few facilities designed for the incineration of PCBs, and the arisings of such materials are so widespread geographically that long distance transportation cannot be avoided. Because the risks of environmental damage are high during the transportation phase, it is essential that the location (nationally and internationally) and transport accessibility should be considered during the planning stages for new treatment plants in order that the optimum location can be determined.

2.0 WORK SAFETY LEGISLATION

2.1 Introduction

This section of the report summarises the principal legislation on health and safety at work applicable to the handling and monitoring of hazardous wastes. It includes a discussion of issues of enforcement and worker participation in the implementation of the legislation.

Workers involved in the generation and handling of hazardous waste inevitably are at a greater risk of possible damage to health, than workers not regularly exposed to, or handling dangerous substances. In addition to the hazards posed by the nature of the substances themselves, hazardous waste workers face additional risks in situations where inadequate or inaccurate information on the composition of the wastes is presented. For example, the receipt of wastes at disposal sites is one occasion when the hazards posed by the wastes to personnel and the environment can be largely unknown. In circumstances where sampling and analysis has been carried out prior to the waste arriving on-site, the undertaking receiving the waste can adopt the necessary safety measures which will largely be determined by the nature of the wastes. Member States, such as Denmark appear to operate this type of system successfully. Similarly, if no prior analysis has taken place, but the documentation describing the load is accurate, the workers handling the wastes again have the opportunity to adopt the appropriate safety measures applicable to the waste in question. This situation is likely to apply in most Member States. However, problems of safety and its level of provision occur when the waste description is either inaccurate (whether deliberately or not) or the hazardous waste load is hidden amongst ordinary waste which does not, by its nature normally present a serious human or environmental hazard. Incidents of this nature are known to occur in virtually all Member States.

A further example of potential hazards to workers is presented by the assessment and treatment of contaminated land. Unless and until a thorough site investigation has been performed, the potential risks to worker health and safety cannot be known with certainty. Undertakings involved in land decontamination work should therefore aim to establish a basic level of operational safety which incorporates a degree of flexibility, perhaps necessitating the provision of additional safety equipment such as radio-telephone moboile first aid unit,etc to safeguard against unforseen hazards.

Despite the potential for adverse health effects, there is no indication at present that workers handling hazardous waste are subject to a greater incidence of ill-health and disease. In many instances, company procedures and the provisions of national worker safety legislation ensure that notice is taken of the enhanced risks and that efforts are made to minimise exposure and reduce any risk.

In the majority of Member States, no specific legislation exists to cover the handling of hazardous wastes. There is, however, extensive 'general' work safety legislation which is used to cover the issue of work safety at handling, treatment and disposal installations. The aims of such legislation, which are applicable to a wide range of industrial and non-industrial activities, are to ensure a safe work environment for all employees and the public. This may encompass aspects of supervision, training, medical care, safety equipment and safety procedures. The enforcement of such legislation is usually the responsibility of a statutory agency. Increasingly, in many countries, a degree of responsibility is placed on both the management and employees concerning the adoption of, and compliance with, safety procedures.

Work safety legislation in some Member States is undergoing a transitional phase as new regulations are being drafted, introduced or enforced. Much of the legislative activity is in response to EC Directives on worker safety. Directive 80/1107/EEC relating to the risks of exposure to chemical, physical and biological agents at work is of particular relevance to the hazardous waste management industry and should lead to major operational changes once implemented and, more importantly, adequately enforced by Member States.

2.2 General Work Safety Legislation

Scope

All Member States have some form of national legislation aimed at ensuring the safety and health of workers. In the majority of Member States the legislation is very comprehensive and the term 'worker' is interpreted in its widest sense. In such cases, the health and safety legislation applies to all employees and not just those involved in manufacturing activities. Thus, the regulations apply equally to workers handling, transporting and treating waste as they do to any other employee engaged in an industrial or commercial activity. Provisions which are common to all national law include requirements for safe working practices, first aid equipment, incident/accident reporting and some level of worker participation and representation.

However, in some cases, the scope of this legislation is restricted by its terms of reference, such that many employees, including those involved in the disposal of hazardous waste are excluded. For example, Irish work safety legislation applies to persons working in 'factories'. Thus, while the provisions of the legislation do apply to facilities, for example carrying out incineration or chemical treatment of wastes, they are not applicable to workers involved in landfill operations or the clean up of a contaminated site. In general terms, it is in fact estimated that up to 80% of workers are not covered by the provisions of the 1955 Factories Act or the 1980 Safety in Industry Act because of the limitations of the definition.

Other aspects of national health and safety legislation effectively limit the scope of safety provision, particularly where companies are relatively small enterprises employing small numbers of people. For example, in France, a legal requirement to set up *Comités d'Hygiène et de Sécurité et des Conditions de Travail* (CHSCT or Committees for Health, Safety and Working Conditions) applies only to companies employing more than 50 people. The Committee has an important role in health and safety matters in terms of investigation, inspection, training and information. In Greece recently adopted health and safety at work legislation will initially only cover enterprises employing more than 50 individuals although it is expected that the provisions will be extended to smaller enterprises in due course.

Medical Screening

Serious gaps in the legislative provision for medical screening exist in the health and safety legislation of Member States. In the UK, for example, no statutory requirement to undertake health surveillance on the workforce exists except in particular industries, such as the nuclear industry. In Italy, where health and safety matters are provided for under two Presidential Decrees, an obligation exists to undertake medical checks on personnel included in any of 57 "processes and categories of worker". However, this provision may not apply to all workers involved in all areas of hazardous waste management and, in Italy there is strong trade union pressure to improve what is regarded as generally inadequate health and safety legislation. In the Netherlands, although most companies have instituted medical checks on employees processing hazardous waste, there does not appear to be a clear legal requirement to do so. Clearly, a lack of routine medical screening of the workforce could lead to a failure to identify any long term risks associated with the handling and disposal of hazardous wastes. Thus, although EC Member States are able to report that physical ill health within the industry is largely limited to accidents involving heavy plant and machinery, the lack of comprehensive health monitoring within Member States prevents a proper assessment of the long-term health risk experienced by workers in the hazardous waste industry.

It is not recommended that health screening should take the place of proper risk assessment and preventive actions. What is proposed is that health screening would provide the evidence that the preventive measures adopted to protect workers health and safety have proved to be unsuccessful.

Safety Officer

Responsibility for safety matters within an enterprise, while covered by national health and safety legislation of most Member States, is a further source of variation between countries. In the Netherlands legislative requirements concern the appointment of a safety officer, either by in-house arrangements or by the employment of an outside safety officer, and the operation of a safety inspectorate and committee, and a worker's conditions committee. In other countries such as Denmark similar arrangement exists, i.e. responsibility for safety matters within an enterprise rests with a safety committee and officers.

In the FRG an 'agent-in-charge' has to be appointed to ensure that plant for the treatment/disposal of special wastes is properly handled and supervised. This is similar to the role of the 'safety officer' found in many industrial plants. The agent in charge, however, has additional powers and responsibilities. These include :-

- the proposing of action to remedy shortcomings in operating procedures. This may involve the recommendation of precautionary measures or technical improvements concerning the disposal of wastes;

- the examination of ways and means of recycling waste materials;

- access to persons with decision-making powers for the plant.

In Spain, where an enterprise is too small to operate a Health and Safety Committee, an individual Safety Officer may be appointed, while in France, total responsibility (and liability) for health and safety matters rests with the head of an enterprise. Thus under the Labour Code and parts of the Social Security Code the head of the enterprise is held fully liable for safety matters and is answerable to the civil and criminal courts. In the civil courts, inexcusable negligence on the part of the employer may increase the compensation payable to the victim of an accident. Articles of the Labour Code place the safety obligation on the head of the enterprise and proven offences may render the offender liable to fines or imprisonment, whether or not an accident has occurred. Proceedings may also be taken against lower levels of management, if the head of the enterprise has delegated powers to persons having the requisite competence, authority and resources.

The head of an establishment is responsible for the application of the regulations that have been imposed, whether or not he is the owner of the premises. Depending on the nature of the enterprise and its activities there are regulations laying down special measures relating to :-

- Classified installations
- Emergency organisation plan
- Monitoring of chemicals
- The disposal of waste
- Environmental protection
- Transport of hazardous substances.

The important aspect here is that liability rests within the head of the enterprise and not just with the company. The extra level of personal accountability thereby provides employers with an increased incentive to ensure that the requirements of the health and safety legislation are implemented. Similar systems of employer/employee and company liability are contained in the worker safety legislation of other EC Member States, e.g. the UK and the Netherlands, where it is a matter for the civil courts.

Worker Participation, Representation and Unions

Articles 5 and 6 of the Protection of Workers Directive 80/1107/EC makes reference to :-

> ".... access by workers and/or their representatives at the place of work to appropriate information to improve their knowledge of the dangers to which they are exposed"
>
> and that
>
> .... workers' and employers' organisations are consulted before the provisions for the implementation of the measures ... are adopted and that workers' representatives in the undertakings or establishments where they exist, can check that such provisions are applied, or can be involved in their application".

These aspects of the Directive have been adopted and encouraged by some Member States and certain industrial enterprises, eg Rechem in the UK, have instigated in-house Worker Councils or Committees to maintain health and safety and make recommendations to improve procedures, even though not legally required to do so. The role of such committees is wide ranging, typically involving discussions on the enforcement and adequacy of existing safety provisions, the analysis of accidents and the development of measures to prevent their recurrence. Such committees are more commonly found within a large treatment or disposal operation or at industrial complexes with on-site treatment capacity.

Greater worker participation in this context is considered to be invaluable as it also reduces the effects of an inadequately functioning enforcement authority because the workers are, to some extent, able to make provision for their own safety and welfare. The trade unions, in particular, have played a major role in seeking to improve work safety conditions in some countries. For example, in Spain the worker safety regulations that were in force during the first half of the 1970s were both vague and general. However, when the General Ordinance on Health and Safety at Work came into force, it was apparent that a number of enterprises with an organised labour movement had already introduced measures that satisfied and occasionally outstripped both the old and new legal requirements.

Enforcement

The enforcement of work safety legislation in Member States is entrusted to an inspectorate, which is operated within, or attached to, the government department of labour, health, industry or the environment. These statutory bodies have wide ranging powers including rights of access and inspection; the prohibition of dangerous practices and enforcement of safety practices.

While variation of duties and powers exists between Member States, in general, their principal responsibilities are similar, i.e. to enforce the relevant legislation and to seek to improve the working environment of all employees.

Though enforcement may be the responsibility of one particular inspectorate, the duties of other enforcement and regulatory agencies may directly or indirectly affect work safety in the context of waste management. For example, in the UK such bodies as Environmental Health Departments, the Mines and Quarries Inspectorate, Police Transport Department, Waste Disposal Departments and Water Authorities all have some input and impact on waste management. The conditions and requirements demanded by these bodies may directly or indirectly influence work safety though this is not generally one of their primary functions.

2.3 Additional Work Safety Regulations

In addition to the limitations and variations in general legislation on health and safety at work noted above, many Member States have made specifc reference to health and safety issues either directly, in their national hazardous waste legislation, or indirectly, as an element of the authorisation procedure for the storage and disposal of hazardous waste. For example, in Luxembourg, waste handling provisions have been incorporated into the general Framework Law on waste disposal, placing certain duties, which are designed to ensure the safety of personnel, on all parties involved in hazardous waste management. Specific reference is made to the provision of first aid equipment and protective clothing. In the UK, the Waste Disposal Authorities, which have the main responsiblity for the operation of the hazardous waste authorisation system, have an obligation to ensure safe working practices and to provide protective clothing for personnel. Similarly, in Spain, authorisations made under regional hazardous waste legislation must make provision for safety aspects of the operation. In the Netherlands, the Labour Inspectorate can make additional regulations on behalf of the Minister.

In Portugal, the 'General Regulations for Health and Safety at Work' states that :-

> "wastes generated in the processing of hazardous or unpleasant substances must be collected and removed as often as necessary, to sites where they cannot cause any danger, and special methods must be used for these operations. Plants specialising in the processing, handling, use or storage of these substances must therefore have facilities for the easy removal of wastes that may be generated. Containers used for wastes, scrap and residues must be constructed in such a way that wastes cannot escape and that they can be easily cleaned. Wastes, scrap and residues must be removed from the workplace so as not to endanger health".

These requirements are still general in that they are subject to interpretation and personal judgement by the enforcement agencies and the waste disposal industry. Thus, they do not guarantee a greater level of worker protection than other work safety legislation which makes no specific reference to safety provision in the handling of hazardous waste.

Important aspects of work safety legislation in the FRG are promoted by the Dangerous Substances Order which aims to promote work processes that reduce the opportunities of personnel coming into contact with dangerous substances, including hazardous wastes, and encourage the use of control measures which are 'state of the art'. Thus, the use of protective equipment, though important, is not regarded as the solution to the problems of ensuring worker health and safety and technical process solutions are viewed more favourably.

The regulations laid down by the FRG are complemented by Accident Prevention Rules (UVVs). They may specify requirements that are either not covered or go beyond those of the state legislation. The UVVs prescribe how an employer has to design installations and premises and what instructions and measures he must take, so as to prevent occupational diseases and accidents at work. There is a UVV on the disposal of refuse which lays down protective measures for use in hazardous waste handling facilities and procedures to minimise the dangers to personnel. However, these do not constitute a comprehensive approach to worker-safety, i.e. they do not aim at technical solutions or intervention in operational procedures.

2.4 Future Legislative Developments

Several Member States are currently in a transitional state, widening applications and extending the measures and requirements of the work safety legislation. These changes are largely in response to a number of EC Directives (82/605, 83/447 and in particlar 80/1107) concerning occupational health hazards and the improvement of the working environment. The aim of the Directives is to reduce the occupational risk from exposure to chemical, physical and biological agents, through a policy of exposure prevention. Frequent medical examinations, worker consultation, provision of emergency measures, record keeping and health surveillance are important aspects of the 80/1107 Directive.

Recently adopted health and safety legislation in Greece incorporates some of the requirements of the Directive. Thus, a new law establishes a local and national framework to supervise the development of health and safety at work. It requires the setting up of company level Health and Safety at Work Committees with worker representation, and specific provisions are made for the protection of workers from harmful physical, chemical and biological agents, all of which are of particular relevance in the management of all aspects of hazardous waste. In Ireland, a new Framework Law on Safety at Work is imminent which aims not only to extend health and safety legislation to all workers but also to incorporate the requirements of the 80/1107/EEC Directive. Likewise, in the UK there are proposals to introduce a new regulation concerning occupational health monitoring and assessment for workers handling or exposed to hazardous substances, thus extending the current provisions. Similar provisions for the protection of employees engaged in work with hazardous substances have been implemented in the Netherlands. The core of the proposed decree will consist of stringent regulations governing the exposure of workers to hazardous substances. It is based on the principle that exposure should be prevented and, where this is not possible, substances must be used which constitute the least possible risk to the health of the worker. If, despite this, unacceptable health risks are present, the work concerned may be subjected to limitations or prohibitions. These measures mirror those of the 80/1107 Directive.

Spain has had regulations governing the examination, diagnosis and assessment of occupational diseases for over 20 years. These orders establish clinical conditions that entitle a worker to welfare compensation and lay down regulations for carrying out medical check-ups prior to, and during employment. Thus, aspects of the 80/1107 Directive have been in place for many years prior to its development. The Spanish regulations have also been modified by more recent provisions to meet compliance with the EEC Directives on Lead and Asbestos.

2.5 Transport and Work Safety Legislation

In some Member States the relevant regulations and requirements concerning work safety during the transport of hazardous waste is contained in separate legislation. However, in common with the general work safety legislation the requirements of the Member States are broadly similar. The following sections will outline the types of regulations currently in force and any future proposals which aim to extend the safety provisions for transport workers.

Within the EC there is no specific legislation specifically concerning work safety during the transport of hazardous waste. Usually the law is more general, aiming to cover the transport of all potentially dangerous substances. The similarities concerning transportation legislation arise because several Member States have adopted the European Convention on the International Carriage of Dangerous Goods by Road (ARD).

It should also be recognised that the law on health and safety of transport workers is enhanced by related legislative provisions. For example, road traffic accident prevention and regulations on working conditions can all contribute to the health and safety of transport workers, though specific reference to hazardous waste transport may not be contained in any single piece of legislation.

Typically, legislation concerning the 'transport of dangerous goods and substances' is the legislation most relevant to worker health and safety during the transport of hazardous wastes. For example, in Belgium, transportation worker protection is contained within the *Règlement Générale pour la Protection du Travail (RGPT).*

The requirements that frequently arise include :-

i) Training for drivers and vehicle operators
ii) Instructions concerning safety procedures in accident situations
iii) Specifications for vehicle construction
iv) The labelling of all packages containing hazardous substances
v) The use of appropriate equipment and precautions.

Though such regulations, which potentially contribute to ensuring worker protection, exist the degree of compliance by industry and their contribution to ensuring worker protection is largely unknown. Similar requirements are found in other Member States where, again, the level of compliance and degree of enforcement is unknown.

2.6 Analysis of Legislation on Health and Safety at Work and the Situation in Practice

In Member States there is a considerable degree of uniformity in the legislative provision in the area of health and safety at work. It is clear that, with few exceptions, the level of provision contained within the national laws of the various Member States is sufficient to safeguard the health and safety of workers involved in the management of hazardous wastes.

While potentially serious limitations do exist, for example, in relation to statutory medical surveillance or limitations to the scope of the law in terms of workers or premises covered, it is understood that future amendments to national laws, largely prompted by EC developments in health and safety legislation, will lead to improved provision in some Member States.

Despite the generally satifactory nature of legal controls, it is clear from the various national reports that existing legal measures are not, in fact, ensuring universally high standards of worker protection throughout the hazardous waste management industry. In some cases deficiencies in the legislation previously discussed, are the obvious cause of poor standards. For example, in Ireland while relatively high standards of health and safety provision exist at establishments incinerating or chemically treating wastes in-house, relatively poor standards may prevail at a contaminated land site.

The majority of cases, however, are examples of poor health and safety practices, which either actively contravene statutory regulations or fall some way short of them. For example, in Belgium, although medical surveillance of the workforce is required by law, deficiencies in the investigation, examination and testing procedures used have been identified. Minimum levels of health and safety provisions exist at one incinerator and one toxic waste treatment centre. In the UK, standards of practice vary across the country and appear to depend, in part, on the level of commitment to health and safety matters shown by the owners/operators of hazardous waste facilities and the resource constraints experienced by the regulatory authorities. It is also clear that in the UK and elsewhere in Europe, the degree of compliance with health and safety regulations during transport of hazardous waste material is unknown. In Portugal, very basic health and safety provision (limited to the use of safety equipment/clothing only) are afforded to workers employed on a landfill site receiving oily and industrial wastes, while relatively high standards of provision are made available to workers in factories (manufacturing pesticides, isocyanates and integrated circuits) where toxic and hazardous wastes are produced, treated and partially disposed of. In the Netherlands some variablility in standards is also apparent in different installations despite the general applicability of national legislation.

Clearly, factors other than legislative provision influence the extent to which satisfactory protection of the workforce is achieved. Some possible factors include :-

- Inadequate information, training and guidance. This has been identified in a number of countries as being a reason for at least some of the bad practices observed at hazardous waste disposal/treatment facilities.

- Differences between the various enterprises in relation to both size and corporate policy. For example, the high standards of worker protection achieved in a Portuguese pesticide factory and a Greek oil refinery are attributed to the inherited health and safety policies of parent companies while similarly stringent health and safety procedures at a British hazardous waste incinerator result from a company policy which is highly committed on safety issues. Differences in the size of company may also account for some variability with larger firms being more willing and able to institute appropriate health and safety measures than smaller companies.

Finally, issues such as the enforcement of the regulations and penalties for non-observance clearly have some influence on the standard of worker protection within Member States. Some difficulty with the enforcement of health and safety legislation has been experienced in some Member States. For example, in Belgium, the management of wastes, setting of conditions and the policing of classified establishments come under regional jurisdiction, whereas powers on health and safety are still in the hands of the central government. This division of power is the cause of many discrepancies between the regions of Walloon and Flanders. The setting up of operational conditions for waste facilities often indirectly and directly affects safety at work and thus the boundaries are not practically defined and conflicts do arise between the regional and central government authorities.

Other difficulties of enforcing health and safety regulations include the problem of ensuring the workforce uses protective clothing, headgear and breathing apparatus. These items are frequently regarded by the users as being cumbersome and unnecessary. Other problems stem from a lack of resources. This appears to be one constraint on the enforcement of health and safety measures, at least in respect of UK hazardous waste installations.

Difficulties with resources can involve both too few inspectors to police a large number of installations as well as a lack of experience in deciding priority action in the absence of statutory guidelines. A national lack of experience may also account for the difficulties faced by some Member States to adequately enforce those health and safety regulations which do exist. Greece is likely to be in this situation for some time to come.

The health and safety at work legislation of many Member States (e.g. Belgium, Netherlands, Spain, France and UK) provides for both fines, imprisonment and closure penalties for infringement of the regulations. However, there is very little evidence as to whether such penalties act as effective deterrants to the contravention of health and safety regulations.

In tables 2.1a - 2.1c the situation, with regard to the provision of health and safety legislation, its scope and implementation, is summarised for individual Member States.

2.7 Conclusions and Recommendations

1. In general terms sufficently comprehensive health and safety legislation exists in all Member States to adequately protect the health and well being of workers operating in the hazardous waste industry.

2. Gaps and limitations are evident in the legislation of some Member States, for example in relation to medical surveillance or limited terms of reference. However, it is expected that future legislative developments and amendments will correct these deficiencies. Any adverse trends, for example in the physical health of workers in the hazardous waste industry, which become apparent as a result of such amendments, should be the subject of study and, if necessary, further action should be taken.

3. Despite the generally comprehensive nature of the health and safety legislation, some variability in the level of provision made exists both between Member States and within Member States.

4. A number of factors have been put forward to explain these differences :-

- inadequate information, training and guidance
- differences in company structure and policies
- enforcement difficulties.

TABLE 2.1a : SUMMARY OF HEALTH AND SAFETY LEGISLATION AND PRACTICE

Provision for :	Health and Safety of Worker	Health and Safety of Public	Defined Employer Duties [1),2),3),4),5)]*	Other Duties and Aspects [6),7),8),9), 10),11),12),13)]**	Future Developments and Comments
Belgium	General legislation not specific to hazardous waste	Refers only to classified plant, some plant evade law	2) Difficult to enforce 3) Subject to Co. policy. Not always carried out	8) No safety committee for firms with less than 50 employees 10) Could be improved 12) Not limiting factor	Division of power between central and regional government poses difficulties
Denmark	General legislation, i.e. not industry specific		1-5) Extensive and detailed at Kommune-kemi and collection stations, through specific regulations	6-10) Generally good, heath and safety enforcement authorities regularly on site, comprehensive monitoring and inspection 11) Satisfactory 12) Not limiting factor 13) Adequate	
France	General legislation covers all workers	Not specified	2) Good 3) Good 4) Good 5) Head of plant has responsibility; may utilise outside expert	6) Achieved through works committee 8) Only for establish-ments with over 50 employees 9) Variable 12) Not limiting	
Ireland	General legislation applies to factories only. Excludes landfill, but		1-3) Satisfactory at incinerator and solvent recovery plant	9) Takes place regularly at specialist treatment	A new frame-work law on safety at work imminent Expected to extend prot-

TABLE 2.1b : SUMMARY OF HEALTH AND SAFETY LEGISLATION AND PRACTICE

Provision for :	Health and Safety of Worker	Health and Safety of Public	Defined Employer Duties [1),2),3),4),5)]*	Other Duties and Aspects [6),7),8),9), 10),11),12),13)]**	Future Developments and Comments
Germany	General legislation covers all workers	Legislative requirements ensure public protection	1-5) Detailed legal provisions covering equipment first aid safety procedures, etc	6) Committees required to examine health and safety issues 9) Satisfactory 10) Satisfactory 11) Given high priority 12) Not a limiting factor	
Greece	New regulations recently introduced	Not specified	1) Very basic at landfill sites, higher at refinery 2) Satisfactory at oil refinery 3) Does not take place at landfill sites 5) An employee is required to ensure safety practice	6-8) Largely absent 9) Poor standards at present 10) Poor at present 11) Unsatisfactory 12) Inadequate 13) Unsatisfactory	Implement-ation not yet satisfactory
Italy	Presidential decrees specific specific to categories of work, including use of toxic materials		3) Takes place at landfill sites regularly 5) At large plants only	9) Duty assigned to ISPEEL (Rome) overlaps with regions 12) Inadequate; provinces responsible 13) Unsatisfactory	Pressure to introduce more safety provisions from trade union organisations

TABLE 2.1c : SUMMARY OF HEALTH AND SAFETY LEGISLATION AND PRACTICE

Provision for :	Health and Safety of Worker	Health and Safety of Public	Defined Employer Duties [1),2),3),4),5)]*	Other Duties and Aspects [6),7),8),9), 10),11),12),13)]**	Future Developments and Comments
Luxembourg	Various worker protection laws and some elements are contained in waste legislation		1-3) Specifically mentioned in waste legislation	13) Unsatisfactory	
Netherlands	General legislation not specific to hazardous waste. Limited at present to premises rather than activity		1) Not always observed, eg at site receiving asbestos 2-3) Satisfactory 5) Either in-house or employ safety expert	7) Not always satisfactory 8) Satisfactory 9) Limited in some cases 10) Satisfactory at many sites 11-12) Inadequate 13) Unsatisfactory	Legislation specifically for works with hazardous material
Portugal	General legislation only		2) Inadequate at landfill sites 3) Satisfactory at some plant eg pesticide plant	6-8) Regular meetings at some plant weekly/fortnightly) eg pesticide plant 10) Unsatisfactory at many sites 11-12) Unsatisfactory 13) Inadequate	New legislation proposed on carcinogens. Standards high at companies with in-house treatment facilities

TABLE 2.1d : SUMMARY OF HEALTH AND SAFETY LEGISLATION AND PRACTICE

Provision for :	Health and Safety of Worker	Health and Safety of Public	Defined Employer Duties [1),2),3),4),5)]*	Other Duties and Aspects [6),7),8),9), 10),11),12),13)]**	Future Developments and Comments
Spain	General labour laws, i.e. non specific		1-3) Depends on size and activity of enterprise 5) Small enterprises without health and safety officer. High standards at some specialist plants	6) Committees in larger enterprises 8) Union represen- tation at large plant 12) Inadequate 13) Unsatisfactory	Transitional stage - improvements in the past owned much to the organised labour movement
UK	General legislation not specific to hazardous waste	Legislation covers public	1-5) Standards vary according to installation and company policy	6-8) Standards vary according to installation and level of awareness. Worker repres- entation is good at some specialist plant 9) Infrequent at most installations 12) Insufficient to monitor installations	Future legislation will require health monitoring of employees handling hazardous substances

KEY :

* Defined Employer Duties

1) First Aid Provision
2) Safety Equipment
3) Health Monitoring
4) Accident Reporting
5) Safety Officer

** Other Duties and Aspects

6) Safety Committees
7) Policy Formation
8) Representaton
9) Monitoring / Inspection
10) Enforcement of Regulations
11) Training of Workers
12) Resources for Enforcement
13) Information / Guidance on Safety Issues

On the basis of these findings the following **recommendations** can be made :-

a) Those powers within the national legislation or regulations which refer to the provision of up-to-date training, information and guidance should be vigorously enforced at all levels.

b) Encouragement should be given to recent trends whereby active participation of the workforce in health and safety issues have resulted in improved safety procedures.

c) A detailed assessment should be made of those organisations and companies which run successful information and training programmes and ways in which such practices could become more universally adopted.

d) There should be no discrimination between workers in the hazardous waste industry on the basis of their company size or structure. Both large and small companies should be required to institute good health and safety practices and to be subject to all the elements of the health and safety requirements. Any difficulties encountered in complying with these requirements should be overcome with appropriate help. Examples might include the provision of training courses on a regional basis catering specifically for small companies. Similarly for those companies which have too few resources to retain specialist medical staff to provide medical care for their workers, arrangements should be made to provide such care on a collective basis, along the lines indicated by Spain.

e) The enforcement of health and safety regulations during the transport of hazardous waste does not appear to be a priority item in the various Member States. Increased awareness through training and education programmes together with increased resources for enforcement should encompass this important and neglected area of health and safety provision.

f) There is a marked reluctance on the part of some workers to wear necessary safety clothing and equipment. Where this occurs, effort should be made to ensure that the equipment provided is as comfortable and unobtrusive as possible, and is of good quality. The provision of suitable clothing and equipment may have cost implications for some companies. The proper use and requirement for protective equipment should be stressed in training and information programmes.

g) Inexperience in the implementation or enforcement of health and safety issues affects few Member States but where it is considered to be an obstacle to the establishment of good safety practices, information and/or technology exchange between Member States should be encouraged.

h) Resource constraints and their effect on the enforcement of health and safety legislation appears to affect only one Member State - the UK, although they are likely to be more widespread than suggested by the national reports. Clearly, the level of provision made by the national government for the purposes of enforcing regulations is a matter of national governmental policy. However, inadequate funding is likely to lead to an increase in adverse health and safety "incidents" resulting in unfavourable international comparisons. More widespread discussion and comparison of health and safety records between Member States may be useful in deterring Member States from providing unacceptably low levels of resources to their regulatory bodies.

i) Further exploration of the role and nature of penalties in the enforcement of health and safety regulations should be considered.

j) More emphasis should be placed on the adoption of preventative measures to reduce worker exposure to hazardous wastes. Both the Netherlands and the FRG actively promote the use of measures such as material substitution, isolation of the worker from direct contact with hazardous waste materials and the discontinued manufacture of some products to achieve reduced worker exposure. It is recommended that the adoption of such measures is more actively promoted in all Member States.

3.0 ENVIRONMENTAL LEGISLATION APPLYING TO HAZARDOUS WASTE DISPOSAL IN THE EUROPEAN COMMUNITY

3.1 Introduction

This section of the report discusses the principal environmental legislation controlling the monitoring and disposal of hazardous wastes. It also introduces aspects of planning control and other related legislation which influence the management and disposal of such wastes.

European Community Legislation

The principal EC Directives of relevance in the field of hazardous waste management are the Framework Directive on Waste, 75/442/EEC and the Directive on Toxic and Dangerous Waste 78/319/EEC. The Framework Directive on Waste places on Member States, a general duty to take the necessary measures to ensure that waste is disposed of without endangering human health and without harming the environment. The terms 'disposal' and 'waste' are defined and certain categories of waste are excluded from the scope of the Directive (e.g. radioactive waste, mining waste, gaseous effluents, wastewaters and some agricultural wastes).

The Directive contains four main mandatory elements :-

- Competent authorities with responsibility for waste are to be appointed
- Waste disposal plans are to be prepared by these competent authorities
- Permits from the competent authorities are to be obtained by installations or undertakings handling waste
- The 'polluter pays' principle is to apply.

Member States are also required to encourage the reduction of waste generation and the recycling and processing of waste for energy generation and raw material recovery.

The competent authorities in a given area are to be responsible for the planning, organisation, authorisation and supervision of all waste disposal operations. The plans drawn up by the competent authorities must cover the type and quantity of waste to be disposed, technical requirements for its safe handling, suitable disposal sites and special arrangements for particular wastes.

Permits or operational licences must be obtained by enterprises which carry out the treatment, storage or disposal of waste. The issuing of such permits by the competent authorities must take into account the type and quantity of waste to be treated, general technical requirements (e.g. hydrogeology of the site, risk posed by the disposal to the local environment) and information concerning the origin, destination and treatment of waste and the type and quantity of such waste.

The competent authorities are required to make periodic inspections to ensure that the conditions of the permits are being fulfilled. Those enterprises that store, treat and dispose of their own waste do not require permits, though they must still be subject to supervision and inspection by the competent authority.

3.2 Toxic and Dangerous Waste Directive

The Directive on Waste 75/442/EEC, laid down the broad structural and administrative control procedures for the control of both household and toxic wastes. Directive 78/319/EEC lays down more stringent controls for toxic and dangerous waste.

The main provisions of this Directive are that toxic and dangerous waste may be stored, treated and/or deposited only by authorised bodies and anyone producing or holding such waste without an appropriate permit must engage an undertaking that is so authorised. Plans are to be made, records kept, transport controlled and inspections made. The Directive also defined toxic and dangerous waste (see section 1.0 above). Member States also have a general duty to ensure that toxic and dangerous waste is disposed of without harming human health or the environment, and in particular, without risk to water, air, soil, plants or animals.

The Directive extends the permit requirement of the Framework Directive to those establishments handling their own toxic and dangerous waste. Any undertaking which produces, holds and/or disposes of toxic and dangerous waste must keep a record of the quantity, nature, physical and chemical characteristics and origin of the waste and the methods and sites used for disposing of it. The transportation of toxic and dangerous substances, whether for treatment, disposal, storage, recovery or recycling must be accompanied by an identification form which details the nature, composition, volume/mass of waste, the name and address of the producer and the name and address of the final disposer and the disposal location.

As with the Framework Directive, all undertakings producing, holding or disposing of toxic and dangerous waste must be subject to inspection and supervision by the competent authorities to ensure compliance with the terms of the authorisation. Member States must also take necessary steps to ensure that toxic and dangerous waste is separated from other matter, appropriately packaged and labelled indicating the nature, composition and quantity of the waste.

In the following sections, the principal aspects of the national legislation of each Member State that directly concern hazardous waste management and environmental protection are discussed. A detailed account of the national legislation is not presented here, this can be found in the National reports. In this section, each country profile comments on the level of compliance with EEC law, indicating the discrepancies or additions that have been made to the national law. Where information is available, an assessment is carried out which examines the effectiveness and implications of the regulations and whether the spirit of the EEC Directive is being upheld.

Aspects of legislation concerning the transport of hazardous waste are discussed separately.

3.2.1 Belgium

National legislation designed to protect the population, workers and the environment from toxic wastes, was drawn up in 1974,the scope of which was determined by the 1976 Royal Decree which defines toxic wastes. Thus, legislation relating to some of the principal objectives of the framework and toxic and dangerous wastes Directives has been in place for over 10 years.

Following the constitutional reforms of 1980, responsibility for the control and administration of hazardous waste rests with the two regions of Walloon and Flanders. The Regional councils, made up of regional members of parliament, have the authority to vote in regional decrees which are legally binding. The co-existence of national laws and regional decrees, the way in which the various concepts are understood and interpreted at national and regional level and the various ways in which the concept of waste itself is defined, makes the legal status of hazardous waste management a particularly complex problem in Belgian law.

The regions are responsible for environmental protection and waste disposal and waste processing. Both regions have decrees which fall broadly in line with the relevant EC waste directives; however, some differences do exist between the regions and certain aspects of the Directives are not covered in national legislation. For example, though the Walloon region has a decree on waste, many of the enacting orders have yet to be introduced and, thus, the actual scope of the legislation is very limited or, in some cases, non-existent. The Flemish region has a similar decree with objectives in line with European legislation and identical to those of the Walloon decree on waste. However, the Flemish outline decree is subject to a number of enacting orders which ensure its effective implementation.

Similarly, the Walloon region has defined major categories of waste that may be subject to special disposal regulations, e.g. 'hazardous wastes' being those wastes which are determined on the basis of their risk to man or the environment. However, no list or criteria has as yet been established by the responsible Executive to determine the boundaries of these definitions. (This may soon be rectified). Waste categories have also been drawn up in the Flemish region and the Flemish Executive is required to draw up a list of hazardous wastes which need special disposal to ensure and maintain human and environmental protection. Thus, the Flemish region is a step ahead of Wallonia in that it has a list (23rd September 1982) of wastes which can be systematically classified.

Measures to promote the prevention, recycling and conversion of toxic and dangerous waste are not covered by national legislation and Flemish orders do not cover wastes intended for recycling. However, both regions have decrees which enpower the executive to regulate or prohibit the manufacture, sale, use of products leading to wastes which are difficult to dispose of or pose a particular threat to the environment. Regional differences in this respect also occur. In the Flemish Region rules can be laid down which prohibit the manufacture, transport, possession and sale of products where certain aspects relating to their longevity and potential for recovery are not satisfied. The equivalent Executive in the Walloon region has no powers to regulate the marketing of products in terms of their longevity and potential for recovery.

Both regions make it compulsory to dispose of all waste and that this should be carried out at the producer' own site or at an authorised facility. This compulsion to dispose is a basic principle unique to Belgium.

Organisation

There are regional differences in the practical aspects of organisation of waste management in Belgium. The Flemish region makes full use of its powers to legislate on and organise waste management policy. The Public Wastes Company of the Flemish Region (OVAM) established by the Flemish Decree of 2 July 1981 has been assigned a comprehensive range of duties relating to all aspects of waste management which may include the construction and operation of disposal plants, the detection of pollution by waste and the compulsory rehabilitation of landfill sites. Thus OVAM has the powers to effectively determine the overall standards of waste management and environmental protection in the Flemish Region.

As already mentioned, enacting orders are still being studied in the Walloon Regions and no structural body has yet been set up to deal with the problem of hazardous waste. This situation encourages traffic in imports and illegal dumping of hazardous waste. Although the current situation is not satisfactory, there are indications that, in the near future, the situation will improve with the adoption of regional orders and the institution of a monitoring authority responsible for closer supervision of existing waste disposal operations.

Administration

Important aspects of the administration of waste management legislation and policy in Belgium are hampered by the slowness of the competent commune and provincial authorities in issuing operating permits, even though a number of statutory time limits are laid down. This situation means that sometimes treatment or disposal operations commence before permits are obtained or are allowed to continue before authorisations are renewed.

The division of responsibility between national and regional bodies also causes problems during the setting up of operating conditions. In Flanders, OVAM determines the technical and procedural requirements of treatment and disposal plant, but no equivalent body yet exists in the Walloon Region. Because operating conditions can include safety aspects, and that safety in the work place remains the responsibility of the Ministry of Employment and Labour, conflicts of power and policy are possible leading to administration difficulties.

Compliance with EC Directives

In addition to the problems outlined above certain other aspects of Belgian legislation as it relates to waste management do not appear to satisfy the requirements of the EC legislation. These include :-

i) the list of toxic substances does not conform to the Annex I of 78/319/EEC;

ii) the obligation to dispose of waste safely is not always achieved in practice;

iii) requirements concerning the labelling of packages are inadequate;

iv) the inadequacy of the obligation to identify wastes being transported, i.e. forms are not required for transport by in-land waterways, sea or air;

v) the inadequacy of the obligation for undertakings which produce, hold or dispose of hazardous waste to supply information to the competent authorities;

vi) disposal sites are not surveyed in all regions;

vii) disposal programmes are not implemented in all regions; and

viii) transportation forms are not required to conform exactly to the information requirements of the Directive. In respect of items iv) and v), Belgium was penalised by the Court of Justice of the European Communities.

3.2.2 Denmark

In Denmark legislation and waste management policy are applied nationally. The fundamental aspect of the legislation is that enterprises must notify the appropriate authority, (i.e. the local council) of the presence and intent to dispose of oil and chemical wastes. These enterprises must then deliver their declared waste to the municipal reception network of transfer stations. Household hazardous wastes can be delivered to municipal collection stations. From these, the wastes are transported to transfer stations. The industrial and hazardous wastes at the transfer stations are then finally sent to *Kommunekemi*, the centralised treatment plant. It is a systematic and well supervised collection and disposal network. Both collection and transfer stations must be authorised.

Enterprises can obtain exemption from this obligation to deliver, provided that they can demonstrate other environmentally acceptable and licenced means of transport and/or disposal.

The regulations applicable in Denmark are based on the principle that anyone who stores, transports or disposes of oil and chemical wastes is held responsible for ensuring that contamination of the physical environment does not take place. A system of fines are, and indeed have been levied if industry does not comply with these principles. Such actions commensurate with a preventative approach to hazardous waste problems. The legislation covers hazardous waste problems as well as contaminated land.

Organisation and Administration

The central authority on waste management is the Environmental Protection Agency. The administration of the regulations lies primarily with the local councils, which can issue orders (bans) and lay down specific regulations to ensure acceptable transportation and storage of oil and chemical wastes.

It is noteworthy that the Danish regulations contain no indication of concentration or quantity limits so that even wastes with a very low concentration of oil and/or chemical wastes are covered by the regulations. This ensures that the producer cannot 'dilute' their way out of the regulation, a problem which is known to occur in other Member States. The Danish definition on oil wastes is also somewhat broader than that employed in Directive 75/439/EEC on waste oils.

Under Danish law empty packaging contaminated with chemical waste is included under the definition of chemical wastes. This has given rise to a number of demarcation problems, since there are no minimum concentration thresholds. In such situations the local councils will judge whether the regulations of the Notice of Chemical Wastes are to be applied. Thus, flexibility is necessary and deemed acceptable if it facilitates safe and practical waste disposal.

A principle of notification is operated whereby enterprises in which the arisings of oil waste are over 150 litres per year or are chemical waste are required to notify the appropriate local council. Thus, local councils are informed of all oil and chemical waste arisings in the municipality. Notification systems operate in other Member States but sometimes these are concerned only with the transport of waste.

Compliance with EC Directives

The obligations of the Framework and Toxic and Dangerous Waste Directives concerning waste description, authorisation, supervision and record keeping etc, are satisfactorily enacted and interpreted in Denmark. Indeed, Denmark could be held up as a model example of how to handle hazardous wastes.

No major proposed revisions of the legislation are foreseen, though a number of adjustments are planned. These aim to make it more difficult to obtain exemption for burning oil waste in small furnaces. In addition, an obligation to notify will be introduced for imported oil wastes and the volume limit on the obligation to notify will be abolished . Municipalities will also be given the opportunity to issue orders to enterprises when there is merely a risk of contamination, thus widening their preventative powers.

3.2.3 France

Legislation encompassing the duties and obligations of the EEC Directive on wastes and Toxic and Dangerous Waste have been drafted by the Ministry of the Environment and implemented nationally. The Waste Disposal Act, 1975, and Registered Plant Act, 1976, already contained the essential aspects of the Toxic and Dangerous Waste Directive and thus compliance has been achieved without major legislative changes.

The regulations are implemented at the level of the basic administration unit, the *Prefecture Departmentale.* This local authority is responsible for the authorisation and supervision procedures on certain classified establishments for the purpose of environmental protection. For such classified establishments, which includes those handling hazardous wastes, a comprehensive system of environmental impact analysis and consultation must take place before an authorisation can be considered.

Regional Variations in Control

National regulations are implemented at local level and this gives rise to some discrepancies. For example, with physical/chemical treatment plants and disposal sites discharging liquid effluent, the quality of the effluent may differ slightly depending on the objectives designed to protect the medium into which the effluent is discharged.

With regard to disposal sites, waste acceptance criteria may differ slightly depending upon the nature of the site (its relative capacity for containment) and operating methods. In general, differences are only minor.

Whereas operating conditions for plants treating and disposing of hazardous waste are fairly uniform nationwide, methods for the rehabilitation of polluted waste dumps and contaminated sites have not been standardised. There are no national or regional standards or directives on methods of treatment or their application, nor are there any standards for the decontamination of sites. Practical decontamination methods are left entirely to the discretion of the local authorities, which generally require the work to be conducted in such a way as to stop the nuisance generated by a tip or a contaminated site without necessarily requiring the site itself to be decontaminated. This practice may be justified technically by the need to use the most appropriate methods having regard to the nature of the problem and the site and the sensitivity of the environment, but its drawback is that it may create technical (and financial) disparities that are sometimes hard to justify. For example, some polluting dumps are dealt with by simple recovery work and piezometric monitoring but then left as they are, whereas other similar dumps are totally decontaminated and the polluting wastes extracted and disposed of.

Recent Legislative Developments and Proposals

The body of regulations specifically covering the management and disposal of hazardous waste were enacted relatively recently, and changes are still taking place. In particular the Seveso Directive on dangerous industrial installations and the monitoring of 'chains of disposal' for nuisance-generating wastes are being implemented. Legislation concerning the laying down of conditions governing the exercise of disposal activities particularly those of carriers of waste and plants treating hazardous waste has not yet been enacted, however, waste treatment actvities can be controlled under the classified establishments law. It has been suggested that reluctance to promulgate the requirement to set conditions arise because this would potentially lessen the liability of producers of hazardous waste - currently regarded as virtually unlimited - by transferring that liability to a statutory approved disposal agency. However, administration delays have also been blamed.

Current Status

There are no major legislative gaps in the national law concerning compliance. That hazardous waste which is disposed or treated at collective installations (1250000 tonnes) is, in most cases, well supervised and controlled by the regulatory authorities. However, some 750000 tonnes of hazardous waste are disposed/treated in-house and it is at these establishments where standards are unknown and may be less than satisfactory. Illegal disposal still occurs but is considered to be extremely limited.

3.2.4 Germany

In Germany the Waste Disposal Act (1975) is the main legislative instrument providing control over "hazardous waste". It contains obligations for authorising production and treatment plants for toxic and dangerous waste and obligations to record, monitor, sample and regulate transport operations. Recent amendments to the Act introduce the concept of waste disposal which includes the recycling and depositing of wastes. The term recycling includes both material and energy recovery. The Act does not specify whether recycling by the owner is considered to be the preferred methodology. With these recent amendments concerning recycling the provisions of the German law are now considered to be compatible with the obligation and duties of the EEC Directives.

Administration and Implementation

The Federal Government provides the legislation framework for the organisation of special waste disposal. In the area of handling and monitoring, however, there is considerable leeway for organisation. It is up to the individual Länder to apply the Federal Law. In some Länder there are private disposal contractors, in others there are compulsory schemes for the disposal of special wastes involving the obligatory use of certain treatment plants.

Many of the Länder have private limited companies which are responsible for disposing or recycling of special wastes. In these Länder, strategic planning has been carried out over a wide geographical area in order to determine the optimum economic and technical solutions that will ensure human and environmental protection. Though they are private limited companies, the regional government usually has a significant holding which may range from 33 to 90%. These companies also frequently operate on a non-profit making basis.

Other regions such as North Rhine Westphalia have a wide variety of public and private companies involved in the treatment and disposal of special waste. The mining and chemicals sector generally have their own incineration plants and tips. The approach in North Rhine Westphalia is based on the expectation that the major producers of special waste will equip and/or expand their own disposal plants. A preference for a private sector solution to special waste disposal is actively sought by some Länder. A private limited company whose shareholders consist of major industrial companies has been formed in Lower Saxony. It has not yet taken over the function of operating the treatment and disposal plants. Such an approach would be useful to monitor, since it is making 'industry' responsible for the disposal of its own wastes. There are advantages and disadvantages to this approach. On the one hand, industry will be able to decide its own protocols for handling wastes that it has most knowledge about. On the other hand, there could be circumstances in which illegal practices are operated. The latter, it is believed, is unlikely to occur.

Saarland and Bremen have considered the construction of chemical-physical treatment plant in their areas of responsibility, but uncertainties over the economic utilisation of such plant resulted in the abandonnement of such plans. Private sector treatment and facilities in neighbouring Länder are used to meet current demands. In summary, the regional responsibilities for implementing Federal legislation, result in a variety of organisational systems, some with a high level of public control, others being operated and controlled by private companies or consortia. However, many regions have conducted a waste management planning process involving co-operation and collaboration with neighbouring Länder. Other important factors include low cost disposal, provision of recycling facilities and the tight control over the operation of all treatment facilities.

The Länder are responsible for the regulations and clean-up of waste and industrial sites which are thought to be hazardous. A legal definition of contaminated land, however, has not been adopted and there are no statutory provisions for the treatment of abandoned sites and contaminated sites in the FRG. Equipment used for the treatment of contaminated soils is subject to approval only if it is a stationary unit. The clean up of contaminated sites will be a key point of the FRG's future environmental policy and technical directives in this areas may be forthcoming.

3.2.5 Greece

The four main legislative provisions covering the disposal of wastes in Greece are :-

- Law 1650 for the Protection of the Environment
- Presidential Decree 72751/3054 on Toxic and Hazardous Waste and PCBs
- Presidential Decree 71560/3453 on Disposal of Mineral Oils
- Presidential Decree July 1986 on Waste.

Law 1650 sets the general legislative background for the protection of all aspects of the environment, covering soil, water, flora and fauna. General provisions are made under the law to allow for the safe disposal of solid waste without deleterious effects on the environment and in such a way as to conserve basic resources.

The Presidential Decree of July 1986 covers many of the requirements of the EEC Framework Directive on waste in relation to its four mandatory elements, that is :-

- The setting up of competent bodies
- Tthe preparation of disposal plans
- The authorisation of disposal activities
- Statistical reporting.

Similarly, the other Presidential Decrees encompasses the main aspects of the Directives on Toxic and Dangerous Waste, Disposal of Waste Oils and PCBs.

Local Directives covering particular circumstances do exist. For example, in the Thessaloniki Prefecture the local directive outlines special conditions which apply to waste and trade effluents, including concentration parameters, sampling and analytical requirements.

In theory, adequate legislation exists in Greece which should not only ensure the safe disposal of toxic and hazardous waste, in addition to making special provision in local difficult situations, but which also brings Greek legislation in line with current EC requirements on waste disposal.

Compliance with EC Directives

The legislation as currently provided, covers all toxic and hazardous waste of concern to the EC to the extent that the Presidential Decree 7275/3054 list of toxic and hazardous materials closely parallels the list provided in the Annex of EEC Directive 78/319.

As far as specifying in detail the treatment a particular waste should receive, or the method of disposal to be followed, the legislation follows closely the line set by the EC Directive. There is a duty of the disposal authority to specify, in the drawing up of plans and in the issuing of licenses, the type and quantities of waste and methods of disposal etc and the encouragement of recycling of toxic and hazardous wastes.

Though the authorities can assume to apportion costs (according the Polluter Pays Principle) under the legislation to finance the future treatment of environmental problems emanating from a waste disposal/treatment facility or for the clean up of contaminated land, there is, as yet, no clear commitment to establish a fund.

A number of local authorities and the Prefecture of Thessaloniki are in the process of drawing up waste disposal plans to include both industrial and household wastes and guidance notes on the handling and disposal of hazardous wastes. The management of hazardous waste in Greece is still in the development stage, with authorities seeking to implement the obligations of the EC Directives. Such action is presently hampered by the lack of appropriate treatment facilities for wastes that provide an alternative to landfilling disposal and the lack of trained staff capable of supervising and inspecting disposal facilities.

3.2.6 Ireland

In Ireland, the Department of the Environment formulates and oversees the implementation of environmental legislation, while the Department of Labour deals with the workplace safety laws. In the absence of a modern framework law on waste management in Ireland the appropriate EC Directives form the main basis of national law. These Directives have been implemented by means of Regulations made under the European Communities Act, 1972. The two key set of Regulations are :-

The European Communities (Waste) Regulations, 1979

The European Communities (Toxic and Dangerous Waste) Regulations, 1982

These Regulations implement, respectively, the Framework Directive on Waste and the Directive on Toxic and Dangerous Waste

The European Communities Act of 1972 does not provide any powers to the Government in using this Act to go beyond legislating for the obligations placed on the State under the relevant EC Directives. The Waste Regulations of 1979 and the Toxic and Dangerous Waste Regulations of 1982 are therefore concerned solely with giving legal effect to the requirements of the corresponding Directives.

The Regulations designate the local authorities responsible for the planning, organisation and supervision of operations for the disposal of toxic and dangerous waste in their areas and the authorisation of the storage, treatment and depositing of such waste. Although the main obligations are, therefore, contained in national law, there is no specific provision with regard to steps to be taken to prevent, reduce, or recycle wastes. Measures to address this discrepancy have not yet been proposed by the Department of the Environment.

Implementation

Responsibility for the authorisation of the storage, treatment or disposal of toxic and dangerous waste rest with local authorities. In 1985 Ireland had some 121 undertakings which might have required authorisations from local authorities; however, at that time, only 10 applications had been received and only 6 authorisations had been issued. Some improvement in the level of authorisation has since been achieved though no data are available.

The use of consignment notes for the transport of hazardous waste has increased significantly during the first two years of operation. The regulation requiring consignment notes was implemented in 1982 and between 1983 and 1984 the number of consignment notes issued by local authorities increased by a factor of 10. As with the number of authorisations, the use of consignment notes is thought to have risen though no data are currently available.

3.2.7 Italy

Presidential Decree DPR915/82, containing regulations implementing the EC directive on toxic and dangerous waste and PCBs, was introduced in Italy in 1982. Before this Decree, the subject of waste management and worker safety was dealt with by general and inadequate health regulations.

This relatively new national framework law has been reached through the process of attaining compliance with Community obligations. This followed a condemnation of Italy by the Court of Justice of the European Communities for failure to comply with the appropriate Directive. Thus, the fundamental principles underlying the 915 Decree concerning authorisation, planning of waste, encouragement of recycling and the 'polluter pays' principle, match those of the waste Directive. The Decree also requires that waste be disposed of according to regional Directives.

Current Status of Regulations in Italy

The regulations supplementing and implementing the 915 Decree on organisation, control and authorisation of disposal services were supposed to be drawn up and issued by the Regions. However, few had achieved this by the initial deadline. Pending the introduction of regional rules to come into force, national regulations are being applied and the regional authorities are enpowered to issue temporary authorisations to undertakings for a limited period of time.

The regulations on recycling came into force from the date of implementation of the Decree and the regions were empowered to promote programmes aimed at limiting the formation of waste, encouraging re-use of waste of and the retrieval of materials and energy sources.

Administration

The co-ordination and drawing-up of general criteria and technical characteristics relating to plans for waste disposal is the responsibility of the national government exercised by the Department of the Environment.

Planning functions are delegated to Regions, which are responsible for identifying areas suitable for siting waste handling and disposal facilities. They are also required to promote suitable initiatives towards waste retrieval.

Control is exercised by the provincial authorities with the assistance and co-operation of local health units or multi-area prevention councils.

Current Status

Regional interpretation of the national regulations concerning the authorisation of commercial dumps varies significantly. For example, some regions have unauthorised dumps operating or issue authorisations without establishing whether the conditions laid down in the regulations are being met. Finally, some regions have adopted the convention of implied authorisation, thus, they do not exercise any advanced control over disposal practices.

The inconsistent criteria for issuing authorisations and the varying degrees of stringency in exercising controls are creating an irregular flow of waste from one region to another. Uncontrolled dumping of waste remains a serious problem in some areas, where no environmental protection measures or effective controls are practical or enforced. However, in other regions such as Lombardy, fairly strict regional laws are operated, for example, relating to the storage of wastes. In Lombardy, temporary storage is allowed only for homogeneous wastes and if these amount to more than 100 tonnes, then authorisation is required.

The problems of putting regulations and planning policies into practice are highlighted in the trends of waste recycling. In 1982 approximately 2.9% of solid waste was sent for recycling, however, since then, the percentage has gone down even further. Other difficulties are likely during the implementation of public controlled waste management policies. This is largely due to the shortage of funds and appropriate communal management bodies.

In summary, there is considerable room for improvement in the management of all waste in Italy before worker, public and environmental protection can be reasonably ensured. Administrative and structural controls need to be adopted uniformly with greater standardisation and consistent implementation of regulations. The transfer of control responsibility to one single body may contribute to the attainment of these objectives. Though national legislation in compliance with the EEC Directives is in place, as yet it has not had a significant or consistent impact on hazardous waste management in Italy.

3.2.8 Luxembourg

The Framework Law (1980) and the Grand-Ducal Regulation (1982) establish two legislative requirements for the planned disposal of toxic and dangerous wastes. The term disposal encompasses the activities of collection, sorting, transportation, treatment, storage and recycling.

The legislation mirrors that of the EC Directives, though some differences do exist. The list of toxic and dangerous substances is more comprehensive than that of the EC Directive, and the Ground-Ducal Regulation lays down concentration thresholds for various chemical substances.

The Grand-Ducal Regulation also requires the holder of toxic and dangerous wastes to describe it very precisely by specifying :-

i) its chemical composition and classification number;

ii) the quantity of waste produced in a month;

iii) the production process and activity of the enterprise from which the wastes originate;

iv) the consistency of the product at 30°C;

v) the special properties of the waste must be described in terms of:

- solubility in water
- nature of any gas that may be emitted
- volatility
- biological decomposition
- internal reaction
- flash point
- pH
- concentration of noxious substances
- water content.

It is noteworthy that the legislation does not exclude from control those wastes with a market value, i.e. wastes intended for recycling.

Concern about the continuing dangers caused by hazardous wastes and the heavy dependence of industry on imports for its supply of raw materials and fuel has led to an overall policy of waste management with emphasis on prevention, recycling and re-use of waste.

A study has recently been completed which examines the feasibility, costs and advantages of such a policy. It should enable the organisation of selective waste collection and the construction of technically controlled landfill sites for the receipt of industrial and difficult wastes.

A national *superdreckskescht* (superbin) scheme is currently being introduced for the collection of particular types of household waste, the early disposal of PCB liquid, closer monitoring of communal dumps and the use of biological monitors to indicate the discharge of wastes to watercourses.

The Regulations also prohibit the intentional addition of water or any other substance to toxic and dangerous wastes before or during collection, the mixing of toxic and dangerous wastes of differing types, sources and properties and the uncontrolled discharge of residues from the processing of toxic and dangerous wastes.

Administration

The Minister for the Environment has general responsibility for the management of waste and the granting of authorisations in particular. Officials of the Ministry have responsibility as regards monitoring and the communes manage the disposal of household wastes. The Minister with responsibility for administration of waters and forests, with jurisdiction over dangerous premises, noxious trades etc, can block authorisations for waste storage, treatment and disposal facilities.

The assignment of certain functions to public authorities (particularly scientific, technical and financial assistance and the stimulation of public awareness) and the clear organisation and allocation of powers at local and national level is recognised as being essential to high standards in waste management.

Monitoring and Implementation

The Framework Law provides that "any breaches of the Law and of its implementing regulations shall be determined and recorded by the criminal investigation department, the gendarmes and the police, and also by the experts and agents to be nominated by a Grand-Ducal Regulation".

Experts include engineers and inspectors from the Environment, Waters and Forests, and Agricultural Administrations and the Factories and Mines Inspectorate. In carrying out their functions, these experts and agents act in the capacity of officers of the criminal investigation department and their jurisdiction extends to the entire Grand Duchy.

In the performance of their task they have access to the premises, sites and means of transport of any person or enterprise subject to the 1980 Framework Law and its implementing regulations and can enter these premises, sites and means of transport during the night when there are serious grounds for presuming a breach of law. They may take samples and items for the purpose of examination or analysis in return for a receipt.

Current Status

Though the adoption of the provisions of Directive 78/319/EEC into Luxembourg law is satisfactory, certain gaps remain. These include :-

- the obligation of the competent authorities to draw up programmes for waste disposal;
- steps to be taken in the event of emergency or serious danger; and
- steps to be taken to encourage prevention and recycling as matters of priority

As mentioned previously, the current development of an overall waste management policy aimed at prevention, recycling and re-use of waste should address the final item listed above.

3.2.9 Netherlands

The Chemical Waste Act (CWA), 1976, aims to prevent the pollution of the environment from chemical waste and used oil. The Act provides the organisational framework which enables effective disposal of such materials and complements the existing legislation (Nuisance Act, Water Pollution Act, Sea Water Pollution Act, Air Pollution Act) in the area of environmental protection. The Act specifies :-

- the methods by which chemical waste may be treated or disposed;
- that a state licencing system (authorisation) must be operated for the storage, treatment or recycling of chemical waste;
- the prohibition of dumping hazardous waste in the ground without an appropriate permit from the Ministry of Housing and Environmental Affairs;
- limitations or prohibitions for the manufacture or sale of goods which, when disposed of, may create environmental problems;
- efficiency treatment methods;
- that yearly plans for the management of chemical waste be drawn up by the Minister.

These aspects are all in accordance with the principles and obligations of the Directive 78/319/EEC.

The treatment of chemical waste and/or its storage on company premises requires no license under the CWA, but is covered by other environmental regulations. Similarly the compulsory transfer of chemical waste from non-licensees to licensees are not covered by the CWA. The transportation of chemical waste is covered by regulations contained in the Hazardous Substances Act.

Implementation

The implementation of the CWA is mainly in the hands of the Minister of Housing, Physical Planning and Environmental Management. The powers include :-

- The recommendation of implementation orders
- The implementation of yearly chemical waste plans
- The assignment of officials to supervise the carrying out of the regulations under the CWA
- The granting of licences and exemptions
- The further implementation of various regulations through ministerial order.

This does not exclude other government bodies playing a role in the treatment and disposal of chemical waste. For example, municipalities, provinces and water boards can make by-laws, provided they do not conflict with the CWA, that may affect hazardous waste management activities.

The officials assigned by the Minister include the Public Health Inspectorate (responsible for monitoring the environment), members of the Control of Hazardous Substances Brigade and Customs and Excise Officers. All these officials check compliance with the CWA by inspection the examination of records and taking of samples.

Various levels of supervision are possible (central, principal, municipal) and the CWA regulates such supervision so that the activities of these bodies can be effectively co-ordinated. The States' Deputies must facilitate consultations between representatives of the various supervisory officials. There is also a provincial co-ordinating Committee for each province which is comprised of representatives of provincial authorities, municipalities, regional environmental inspectorates, the Controllers of the Hazardous Substances Brigade and the Customs and Excise Inspectorate.

Current Status

Under the present CWA system, chemical waste is designated in accordance with the concentration of certain toxic and dangerous substances. This system can lead to problems.

Firstly, the lowest concentration level under the CWA (50mg/kg) is probably too high for some extremely toxic substances (e.g. mercury). Secondly, laying down the lowest concentration levels allows for the mixing of various (waste) substances, which results in the making of a non-chemical waste (under CWA definition) out of chemical waste. This system has been under review and new legislation will be introduced in 1988. Thus chemical waste will be defined :-

- by a chemical waste list (naming some 1000 waste streams);
- in accordance with the concentration of certain toxic and dangerous substances within these waste streams; and
- by new (lower) concentration thresholds of very toxic substances.

Recent policy in regard to malpractice in handling oil waste has, therefore, been more stringent. In particular, it is aimed at preventing the mixture of used oil, or fuel made from this, with polycarbons or organic halogens. Two decrees have been promulgated for this purpose, one of which gives a clear definition of used oil and of what is to be understood by chemical waste. Both decrees are unusual in that they specify the methods of analysis for the measurement and calculation of polycarbon and organic halogen contents and also the method of sampling.

Proper legal provisions are lacking at present in relation to representative sampling and analysis of waste. The absence of clear guidance on these matters has meant that the sample and its analysis have not been acceptable as sufficient evidence in the prosecution of offenders under the Chemical Waste Act and, as a result, many cases of this nature may be dismissed. This makes the enforcement of the Act difficult. At the request of the Inspectorates of Health and Environmental Protection of the Ministry of Housing, Physical Planning and Environmental Management, research has been carried out to establish guidelines for the sampling and analysis of (chemical) waste materials. The intention is to incorporate the results in new legislation.

A waste producer or waste owner may decide for himself whether to keep such waste or to dispose of it by transferring it to third parties explicitly designated by the Act. The licencing requirement of the Chemical Waste Act does not apply to chemical waste which the producer himself treats or stores on his own property. This approach was adopted in order to stimulate waste recycling by the producers themselves, and at present approximately 50% is treated in-house and approximately 20% is recycled.

In view of the large quantities (ca. 450,000 tonnes) per year of chemical waste which are thus excluded from the scope of the CWA, it is difficult for the government to gain insight into the whole problem of chemical waste. The fact that under other environmental legislation conditions may be imposed on the storage and treatment of chemical waste, i.e. through the Nuisance Act, does not provide an immediate solution to this problem.

A further aspect is that the CWA scarcely makes any provision for instruments which can be used on the preventative front. The preventative approach should be aimed specifically at non-treatable chemical waste. At present non-treatable chemical waste can be landfilled under strict conditions and, thus, the pressure to take preventative measures is reduced. If a preventative aim as well as a criterion for effectiveness were to be added to the preamble of the CWA, for example, this might give some stimulus towards a more preventative policy. But, on the other hand, the enforcement of the Act in practice causes problems. One of the reasons may be the centralistic structure of the Act. Improved co-ordination at the various levels could probably help to ensure adequate supervision of compliance.

The market value of chemical waste is bound up intrinsically with many of the problems of enforcing the regulations laid down by the CWA. Some possible solutions are :-

- to make import and export regulations more stringent;

- the introduction a system of passing on the costs, which would mean that the illegal transfer of chemical waste no longer has any advantage; a regulation in the area of taxation combined with an exemption system for the use of certain raw materials is one of the possibilities here; or

- the collection of chemical waste to be the responsibility of the public authorities.

3.2.10 Portugal

The legal framework governing the management of toxic and dangerous wastes and environmental management in general has been introduced only relatively recently in Portugal. The most recent legislative actions on waste have been prompted by Portugal's recent accession to the EC and the need to comply with Community Directives on the environment.

Thus, there are a number of decree laws which correspond to the obligations and principles of the EC directives on waste materials (i.e. 75/442, 78/319, 75/439, 85/467). Several other laws on the environment have already been prepared, based mainly on relevant Community directives, e.g. quality of surface waters, waters for human consumption, bathing waters; classification packaging and labelling of hazardous substances; quality of industrial effluents. However, some of these have not yet been implemented and the Directive 78/319/EEC on Toxic and Dangerous Waste does not have to be ratified until 1st January 1989.

Assessment of Current Situation

The most common method of disposing of toxic wastes in Portugal is by uncontrolled tipping (78%). In this context 'uncontrolled tipping' is considered to occur where one or more of the following conditions have not been fulfilled :-

- the site is lined with impermeable material;
- leachates and surface waters should be drained and controlled;
- contaminated water should be treated.

Wastes disposed of in this way are primarily organic and inorganic solids and sludges generated by production processes and, in some cases, by effluent purification plants. For example, pyrite residues contaminated with arsenic and heavy metals generated during sulphuric acid manufacture, are landfilled in large volumes without environmental safeguards.

Approximately 4% of all wastes generated in Portugal undergo recovery and re-use. Major examples being oil wastes, which are used as an energy source and solvents, which are recycled. Uncontrolled tipping takes place both on sites owned by the waste producers themselves and on municipal sites. Monitoring of waste arriving at municipal sites is not common practice.

Regulations on industrial liquid effluents have not yet been introduced and at present there is no legal control over liquid discharges of existing industrial plant. In these circumstances toxic and dangerous substances are often discharged in liquid form to sewers, water courses or even directly to the soil. A strict implementation of the future legislation on industrial liquid effluents will result in a substantial increase in the arisings of hazardous waste.

In summary, though the general waste legislation has been implemented, Portugal still has no adequate system for monitoring and controlling the production, storage, treatment and disposal of hazardous waste. Industry within Portugal is not yet adequately informed of the methods they must adopt to comply with the new laws and uncontrolled tipping is likely to prevail as the main disposal route for hazardous waste.

3.2.11 Spain

The national law follows the precepts of the EC Directive on Toxic and Dangerous Wastes, however, there are some aspects which differ (e.g. waste definitions).

Before giving authorisation for the setting up of an industry that generates toxic and dangerous wastes, the Spanish Government may require the applicant to take out civil liability insurance to cover any liabilities that may result from his activity. Producers are required, without prior request, to submit an annual report stating the quantities of waste generated or imported, detailing their nature and final destination, and to notify the authorities immediately in the event of the disappearance, loss or escape of toxic or dangerous wastes.

For authorisation to carry on waste management activities, the applicant is required to take out civil liability insurance and pay a bond in the form and for the amount laid down in the law. The applicant must also establish safety measures, internal emergency plans for alarm system evacuation procedures and assistance arrangements with the emergency services.

The Basic Law on Toxic and Dangerous Wastes also makes provision for the creation of waste exchanges. This aspect and that concerned with liability and penalties are not paralleled in the Directive.

The enforcing regulations have not yet been introduced but final drafts indicate that they will include clarification of definitions, the requirements for safety provisions, civil liability insurance, aspects of storage, transfer centres and the control over the flow of wastes to treatment plants.

Several Articles of the Draft Regulations are dedicated to the legal provisions covering small and occasional producers of hazardous waste, as the usual provisions are impracticable when applied to enterprises that generate waste irregularly or in small quantities. Such enterprises will be exempt from certain obligations.

Regional Legislation

Although central government is responsible for basic legislation and has the power to adapt Spanish law to bring it into line with EC Directives, it is the Autonomous Communities that are responsible for implementing national legislation in this area.

In short, central government has the responsibility for laying down minimum basic regulations and co-ordinating the actions of the Autonomous Communities. These Communities are empowered to issue additional regulations on protection and enforcement of the basic national legislation and will also be responsible for implementing the laws and authorising and monitoring activities. Municipal Councils, too, have responsibilities in conjunction with, or instead, of the Autonomous Communities for the collection of wastes in certain cases and for authorising and monitoring waste-generating industries.

Certain differences between the Catalan Autonomous Community Law and National Law are apparent. For example, though both comply with the toxic and dangerous waste definition of the 78/319/EEC Directive, they are not identical and the types of waste excluded from the scope of the regulations are also slightly different.

The draft regulations of the Community of Madrid also differ in some respects from national law in that the concept and, thus, the definition of toxic and dangerous waste is much broader.

Current Status

Pressure of time led the Government to pass a basic law on toxic and dangerous wastes five months after accession to the EC, but the regulations merely transpose the EC Directive into Spanish Law, without any knowledge of the real situation. A similar situation applies in the case of contaminated land where studies and action programmes have been launched, but no general co-ordinated action has been taken to locate contaminated land, assess the damage and restorative measures.

The problems created by the absence of a national programme to implement the law are compounded by the very wide disposal powers, which tends to diminish the effectiveness of the administrative bodies. There are five separate ministries with powers regarding toxic and dangerous wastes and each of these has several departments performing specific functions.

It should be borne in mind that 14 of the 17 Autonomous Communities use Central Government as a model for organising their own environmental departments, which means that numerous bodies are replicated throughout the country. This structure does not enhance the effectiveness of the control authorities and makes co-ordinated action difficult to achieve. In addition, divisions of power between Central Government and the Autonomous Communities remain unclear.

The Enforcing Regulations of the national law which will probably be passed in 1988 should lead to some improvement in the management of wastes in Spain, which, at present, is far from satisfactory. They will also fill some of the existing gaps in the national legislation relating to making special provision for waste oils.

3.2.12 UK

The main elements of the Framework Directive on waste (75/442/EEC) have been satisfied in England, Wales and Scotland largely by the Control of Pollution Act (COPA), 1974 and in Northern Ireland by the Pollution Control and Local Government (NI) Order, 1978 although other legislation, such as the Local Government Acts, are put forward in support of compliance in some areas of legislation.

In some respects, the COPA legislation is more exacting than the Framework Directive in that licensing provisions apply to undertakings disposing of their own waste as well as being disposed of for third parties. In addition, the COPA requires publication of waste disposal plans and a period of consultation with other bodies during the preparation of the plans.

Requirements made under the Directives dealing with Toxic and Hazardous Waste (78/319/EEC) PCBs and Mineral Oils are, for most part, satisfied in UK legislation by the Control of Pollution (Special Waste) Regulation, 1980, although there are some areas in which true compliance has not been achieved.

These concern the issue of recyclable toxic and hazardous wastes (covered by the Directive but not by the UK legislation); the lack of a requirement to keep records on toxic waste at production and storage stages in the UK legislation, in addition to the transport stages and at the point of disposal; the exemption of wastes produced by the Crown under UK legislation but not under the Directive.

The principal exception is the Directive on PCB disposal which makes collection and/or destruction and regeneration of waste PCB a compulsory process. Under UK legislation, there is no compulsion to dispose of waste PCBs in a prescribed manner. Thus, an individual could hold a quantity of waste PCBs on his premises until an occupational hazard is posed.

Current Status

The setting and enforcement of licence conditions for waste handling and disposal facilities is central to waste management in the UK. The conditions set by the local Waste Disposal authority takes into account local environmental conditions and hydrogeology. Thus, it is normal practice for a full site investigation to be required by the local Authority prior to the issuing of a licence of granting or planning permission.

Proper enforcement of appropriate site licence conditions is fundamental to the attainment of environmentally adequate, equitable and consistent standards of hazardous waste management. This is not uniformly achieved in the UK.

Inspections of waste facilities are usually carried out unannounced and typically involve a visual inspection as to the receipt of wastes, disposal methods employed, security and tidiness of the site. Though Waste Disposal Authorities (WDA) have a duty to enforce licence conditions on private and public authority sites, there are no statutory requirements concerning the frequency of inspection. It remains the discretion of the WDA officer to determine the frequency and detail of inspections.

One of the practical problems associated with inspection is that with the infrequency of visits improper disposal can take place unnoticed and there is little the inspector can do to rectify a potential hazard once unsatisfactory disposal has occurred. Therefore, there is considerable reliance on the part of the operator of waste disposal facility to comply with the licence conditions and ensure that he operates the facility with due regard for the safety of the environment, public and workers. In the great majority of cases there are few problems and where pollution incidents from environmental contamination has occurred, detailed investigations have taken place to determine the cause and appropriate action has been taken to rectify the problem.

The judgement and experience of individual WDA officers will also influence the inspection procedure and assessment of compliance. However, in some WDAs the nature and frequency of inspection is considered to be unsatisfactory because the authorities do not have sufficient trained personnel to carry them out. That is, frequent and adequate inspection is not practicably possible. It has been proposed, by some WDAs, that frequent inspection of facilities should become a duty of WDAs in order to improve standards in these areas.

The whole subject of the adequacy of the site licencing provisions of COPA has recently been reviewed by the Control of Pollution (Special Waste) Regulations Joint Review Committee. The Committee reported that the site licencing provisions of COPA embraced a comprehensive code of law which they considered to be sound in general but with 'certain weaknesses'. For example, the committee reported that consideration should be given to extending the scope of the licence conditions to cover the post-operational period of the site. Secondly, that the Department of the Environment should bring forward proposals for regulations to prescribe specific operational methods for waste disposal and the related use of plant and equipment. This would strengthen the control achieved through site licencing itself.

There are numerous other aspects besides site licencing, which have been identified as contributing to hazardous waste management problems in the UK. For example, the Hazardous Waste Inspectorate expressed concern that Government advice and trade association Codes of Practice were being disregarded, wastes were being deliberately mixed and standards of enforcement showed considerable regional variation.

Other factors that diminish the effectiveness of hazardous waste control include :-

- The long time delays associated with court proceedings involving breaches of licence conditions

- The lack of up-to-date guidance on 'best practice' in the disposal of specific types of waste and the setting of licence conditions

- The inadequacy of the current 'special waste' definition and its lack of regard to environmental protection

- Commercial incentives to hide hazardous waste and thereby reduce disposal costs.

Certain proposals have been drafted which, if introduced, would extend existing waste legislation in four main areas :-

- The introduction of a 'duty of care' of waste producers (i.e. to ensure the satisfactory disposal of their wastes)

- A registration scheme for all waste handlers

- New powers for waste disposal authorities to set restoration and aftercare conditions for landfill wastes

- To define storage and thereby require operators of industrial premises to obtain an authorisation from the WDA when they are storing waste.

Apart from the complexity of translating each of the above proposals into workable legislation, opinion within different sections of the waste disposal industry is deeply divided as how such proposals should be introduced. It is likely, therefore, that there will be some considerable delay before any new legislation on the disposal of wastes is introduced in the UK.

In tables 3.1a and 3.1b the legislation and practice relating to hazardous waste within industrial Member States is summarised.

3.3 Planning Controls

In addition to compliance with EC Directives and specific national regulations on waste, control over hazardous waste can also be exercised through planning or building procedures. These procedures may require that an environmental impact assessment of a waste disposal or treatment facility is carried out whereby the natural environment and socio-economic impacts of the facility are considered in relation to the existing environmental conditions and regional character of an area. The depth and emphasis of such an analysis can vary, but typically the construction of a facility, its operation, emissions, proximity to residential and amenity areas, access arrangements and public health aspects are all considered. In the UK this process is achieved through consultations between a planning authority and expert bodies responsible for ensuring water quality, public health and worker safety. Through this process, it is possible for the planning authority to make the approval of the development conditional upon the adoption of specific operating or plant design criteria. These conditions can involve provisions to protect the environment, local amenity and public health. Similarly in France, an impact study is required on classified establishments before they are granted an operating authorisation. The results of such a study are investigated by the *Inspection des Etablissements Classées* and submitted to the *Conseil Départmental d'Hygiene* for its views.

This impact analysis/planning process varies within the EC and in some Member States it has only recently been introduced. Thus, there are numerous specialist waste treatment plants and landfill operations which predate planning and site authorisation procedures, and thus, have never been fully assessed with regard to their suitability as sites for the receipt of hazardous (or non-hazardous) waste.

TABLE 3.1a : ENVIRONMENTAL LEGISLATION AND PRACTICE

	Belgium	Denmark	France	Germany	Greece	Ireland
Legal Definition of Haz. Waste	/ Regional discrepancy	/	/	/ Precise definition	/	/
Competent Bodies - Effectiveness	/ Poor	/ Good	/ Variable	/ Good	/ Good	/ Variable
Waste Disposal Plans of Competent Bodies	■	/ No oblig-ation to draw up plans	■	/	■	/ Incomplete submission
Permits for Facilities	/ Not always considered	/	/ Conditions not set	/	/	/
Inspection of Facilties	/ Poor	/ Good	/ Variable	/ Good	/ Poor	/ Variable
Record Keeping						
- Storage	■	/	/	/	/	/
- Transport	■	/	/	/	/	/
- Disposal	■	/	/	/	/	/
Situation Reports	■	/	■	/	■	■
Transport Documents	/ Inadequate	/	/	/	/	/ Partial compliance
Recycling of Waste	■	■ Recent legislation	/	/ Legal obligation	■ No legal obligation	■ No legal obligation
Liability Insurance	/	■	■	■	■	■
Overall Status of Legislation and Enforcement	Improvement needed	Good	Improvement needed in some areas	Good	Poor but some improvements taking place	

■ No legislative provision/conditions not fulfilled

TABLE 3.1b : ENVIRONMENTAL LEGISLATION AND PRACTICE

	Italy	Luxembourg	Netherlands	Portugal	Spain	UK
Legal Definition of Haz. Waste	/	/	/	/	/	/ No environ environmental aspects
Competent Bodies - Effectiveness	/ Poor	/ Variable	/ Variable	/ Poor	/ Poor	/ Variable
Waste Disposal Plans of Competent Bodiest	■	■ No legal provision	/	■ No plans submitted	■ No plans submitted	/ Many not submitted
Permits for Facilities	/	/	/	/	/	/ Not always satisfactory
Inspection of Facilties	/ Poor	/ Variable	/ Variable	/ Poor	/ Poor	/ Poor
Record Keeping						
- Storage	/	/	/	/	/	■
- Transport	/	/	/	/	/	/
- Disposal	/	/	/	/	/	/
- Comment	Poor	Variable	Variable	Poor	Poor	Poor
Situation Reports	■	■	■	/	■	■
Transport Documents	/	/	/	/	/	/
Recycling of Waste	/	■ No legal provisions	■	■	/ Waste exchange to be set up	■ No legal provisions
Liability Insurance	■	■	■	■	/ Compulsory	■
Overall Status of Legislation and Enforcement	Transitional and improvement needed			Transi-tional impro-vement needed	Transi-tional impro-vement needed	Improvement needed in some areas

■ No legislative provision/conditions not fulfilled

With the introduction of the Environmental Impact Assessment Directive, 85/337/EEC, there will be a requirement that all new waste disposal installations for the incineration, chemical treatment or landfill of toxic and dangerous waste shall be subject to an assessment. This will require that the development of such installations pays due regard to the project size, design and environmental impacts and will ensure that appropriate measures are taken to reduce these impacts. Effective implementation of this Directive will substantially improve the planning and site authorisation procedures of some Member States and may contribute to the greater harmonisation of procedures throughout the EC.

In the FRG, plans for the construction and/or operation of stationary plants for the treatment of special wastes, or any significant modifications to such a plant, have to be vetted by the competent authority. However, the details of the vetting process are stipulated by the Waste Disposal Act and the Federal Air Act, and not, as in the UK, in planning regulations. As yet there are no generally recognised and harmonised directives and references that could be used in the technical planning of such plants. In practice, wide discretionary powers exist in respect of the application of the approval procedure and it may take 2 to 10 years to get such a plant approved. A Technical Directive on Waste, which is currently in the pipeline in FRG, may go some way to reducing the time delays associated with planning and authorisation permits and eventually may enable the harmonisation of plant design and technical procedures.

It should also be noted that construction and authorisation permits in the EC are usually only applicable to stationary treatment facilities and that mobile units are not subject to the same statutory approval and administrative delays. In FRG there is a discernable and marked emphasis on the use of mobile and semi-mobile equipment particularly for the treatment of contaminated soils.

Throughout the EC, any facilities handling, treating and disposing of hazardous waste have the potential to cause serious environmental damage. In this respect an effective planning and plant authorisation system can be invaluable as a means of predicting and mitigating such impacts. Some Member States, (e.g. France, Germany, UK) successfully operate such systems, however, the practice of uncontrolled tipping and the operation of unsuitable landfill sites (e.g Portugal, Greece) suggests that some Member States would benefit considerably from more stringent planning and authorisation procedures.

3.4 Transportation of Hazardous Waste

The legislation concerning the transportation of hazardous waste in the EC is governed primarily by the relevant articles of the Directive on Toxic and Dangerous Waste; national regulations concerning the transport of dangerous substances and certain international agreements on road and rail transport.

In the section below the relevant aspects of hazardous waste transport legislation are broadly outlined for each Member State, indicating the difference between national and EEC law and the degree of implementation and enforcement.

3.4.1 Belgium

The transportation of wastes in Belgium does not require a specific authorisation though there are requirements which carriers must satisfy. If the carrier is the acquirer or importer of toxic waste, then he may be obliged to obtain an authorisation accompanied by an undertaking to take out an insurance policy.

The Walloon decree on wastes empowers the Executive to make the transportation of certain wastes subject to authorisation but there is no enacting order to enforce this regulation. The Flemish laws do not contain any authorisation system for transportation per se, however, the removal of wastes is subject to prior authorisation. In addition, any delivery of wastes including operations by which wastes are transferred to carriers must be declared.

Under the Royal Decree of 13 January 1986, bringing the 'ADR' (European Agreement concerning the International Carriage of Dangerous Goods by Road) regulations into force, an' ADR' certificate issued by the approved bodies for vehicle inspection must be carried on board all vehicles which transport hazardous wastes in fixed or transportable tanks or in tanker containers of more than 3,000 litres. Transport documents must accompany any transport and include the ADR conformity certificate.

In addition to the above regulations, there are technical regulations applying to tipper trucks used for carrying hazardous waste and general regulations applying to all automobile vehicles. The Royal Decree implementing the ADR is supplemented by some Ministerial Orders approving inspection bodies for the checks, tests and trials required for the transportation of hazardous waste by road.

3.4.2 Denmark

Transportation of hazardous waste in Denmark is regulated in accordance to Notice No 2 of 2nd January 1985 on the National Transportation by Road of Dangerous Goods. Under the Notice on Chemical Wastes, a consignment of chemical wastes must be accompanied by a special chemical wastes card. These cards give directions on the nature, packaging, labelling and transportation of the wastes, together with action in the event of an accident.

Transportation by rail of oil and chemical waste must follow general regulations for the transportation of dangerous goods. Transportation by sea observes the rules of the Government Shipping Inspectorate and the Ministry of Industry's regulations.

3.4.3 France

The regulations on the transport of hazardous substances in France cover internal transportation by land, rail and inland waterways. A number of provisions of the *Code de la Route* directly relate to the transport of hazardous substances for example, they include provisions covering :-

- Technical inspection of vehicles
- Prohibitions on the movement of hazardous substances
- The use of the public highways for the movement of hazardous substances.

The international transport by rail and road are subject to controls and requirements of ARD and RID (Regulations concerning the International Carriage of Dangerous Goods by Rail).

Under the *Réglémentation Rélative au Transport des Matières Dangereuses*, there are requirements covering :-

i) classification of substances, loading and unloading precautions, information on data sheets;

ii) a review of the 14 classes of hazardous substances;

iii) labelling, marking of vehicles, driver training;

iv) administrative regulations; and

v) nomenclature of substances.

Vehicles may be required to carry special equipment, e.g. extractors, appropriate fire-extinguishing agents and hoses. Drivers of such vehicles carrying hazardous substances must have attended a training course and obtained a training certificate from a specialist body. Training is compulsory for transport of tank trailers and road tankers as well as for radioactive substances and explosives.

Thus, in French national law there are detailed requirements concerning the transport of hazardous substances that should contribute to the reduction of worker, public and environmental hazards. However, information concerning the effectiveness and the degree of implementation and compliance of these regulations is not readily available.

3.4.4 Germany

In line with the 78/319/EEC Directive, the Waste Disposal Act prescribes that special wastes may not be collected or transported without a transport permit. The following aspects are also considered before a permit is issued:

- Reliability of carrier
- Vehicle suitability
- Compliance with Dangerous Goods Regulations
- Disposal will be at an authorised premises.

Transport permits for special wastes are valid only for a given period (usually two years). The supervising authorities can therefore maintain control in conjunction with the obligatory technical checks on vehicles.

A Federal Government order on the importation of wastes requires that all entries on application forms for the importation of wastes have to be in German. This facilitates the assessment of criteria used to define the waste. Also the exporting or forwarding of waste out of FRG requires a special shipping ports.

3.4.5 Greece

The transport of waste materials in Greece is, in part, controlled by the international agreements in the transport of dangerous goods, (ARD and RID) and the requirements of the 78/319/EEC Directive. The degree of compliance and enforcement of the regulations is unknown.

3.4.6 Ireland

In addition to the general transport legislation applicable to all road vehicles, the EC Toxic and Dangerous Waste Regulations, 1982, contain certain provisions relating to the safe transport of hazardous wastes. These mirror the relevant Articles concerning the transport of wastes contained in the Directive 78/319/EEC. They include the appropriate labelling of wastes and the operation of a consignment note system, where information concerning the nature, quantity, producer and destruction method of the waste load are specified.

The extent of compliance with the labelling and storage requirements is unknown. Information on the operation of the consignment note system is only available for the year 1985. In this year, only half of the local authorities had implemented the system. However, over the three years following enactment of the regulations in 1982, the trend was towards increasing implementation.

3.4.7 Italy

There are no regulations specific to the transport of hazardous waste in Italy, other than those that implement the relevent Articles of the Directive 78/319/EEC.

3.4.8 Luxembourg

The Framework Law provides that anyone collecting and transporting wastes by way of trade must have prior authorisation, which will be granted only if the operations are carried out without endangering human health or the environment in general. Conditions relating to technical equipment may be attached to the authorisation. Thus, the requirements are in line with the relevant Articles of Directive 78/319/EEC.

3.4.9 Netherlands

The Chemical Waste Act, which is the main legislative act concerning hazardous waste, makes an exemption provision for the transportation, packaging or supply of chemical wastes when transferred to third parties. However, the transport of hazardous materials is regulated primarily by the Hazardous Substances Act. The Act specifies eight categories of hazardous materials which are divided into a number of sub-classes. In general, these classes consist of materials which are either explosive, pressurised, inflammable, toxic, radioactive or corrosive. Thus the inclusion of chemical waste under the requirements of the Act depends on it possessing one of the above properties. Chemical wastes that are within the scope of the Act are then subject to a set of detailed regulations relating to :-

- Packaging, documentation and loading
- Testing
- Danger labels
- Vehicle construction, tanks, technical equipment and inspections.

Danger and substance identification numbers, transport documents and instructions for dealing with accidents must be used by carriers. However, the regulations of the Hazardous Substances Act cover pure substances or combinations of substances which have certain properties. Chemical wastes, however, are usually impure mixtures and difficult to define under the terms of the regulations. Thus certain chemical wastes may obtain exemptions regarding the requirements of transport and packaging because of the limited scope of the regulations.

Enforcement of the Hazardous Substances Act in the Netherlands is the task of the Commandant and Controllers of Dangerous Substances Brigade.

In comparison with the terms of the Directive 78/319/EEC, in relation to the transport of chemical waste, the following aspects are important :-

i) The producer of chemical waste is not obliged to fill in a form to accompany transportation.

ii) The producer may only hand over chemical waste to a carrier who holds a transport licence.

iii) The carrier is not obliged to keep a system of documentation and/or supply any information on request.

The producer of chemical wastes is confronted by a whole range of statutory obligations. However, many of the regulations and requirements are intended to promote the safety of commercial and industrial activities involving chemical products (often with clearly defined properties). The operation of three different systems of indicating danger (Transport of Dangerous Goods by Road, Delivery of Dangerous Goods, and Safety Regulations) is, especially for mixed loads of very different types of waste, most impracticable.

The regulations concerning the delivery of dangerous goods is designed to give protection against goods which are placed on the market. This is not the case with chemical wastes, thus it may be appropriate that chemical wastes should be exempt from non-relevant statutory regulations as adequate control and coverage would appear to be guaranteed by the Safety Alert Decree and by the implementation of the Overland Transport of Dangerous Wastes Decree.

3.4.10 Portugal

Portugal has sought to establish national legislation in compliance with the EC Directives. The Governmental Decree 374/87 corresponds to Directives 75/442/EEC and 78/319/EEC, however, Portugal has derogated ratification of the Directive on Toxic and Dangerous Wastes until 1st January 1989, a Directive which includes certain transport requirements.

Information concerning the implementation of the Governmental Decree is not available. However, in general, wastes are rarely declared as toxic or dangerous by the industries involved and it is doubtful that adequate precautions with regard to safety procedures and equipment are taken during their transportation.

3.4.11 Spain

The National law states that operations must be accompanied by a special waste-identification document which is required by the 78/319/EEC Directive. However, details are left to the Enforcing Regulation which should be promulgated in 1988.

Within the draft (Enforcing) Regulations the haulier may refuse to accept a load if the characteristics, nature and quantity of the wastes transported, do not match those recorded by the producer on the consignment note. In this event, the haulier is obliged to notify the competent environmental authorities.

Waste transporters must take out insurance to cover any damages that may be caused by the wastes from the time of loading to the moment they are delivered to the manager (the compulsory insurance requirement is contained in the Transport regulations and not within enforcing the new Regulations).

Other aspects are governed by the regulations on the transportation of dangerous goods by road, rail, sea or air.

Regional differences, with regard to hazardous waste transport legislation, do exist, though the essential points are in-line with the National Law. Under Catalan Law wastes have to be accepted by the treatment plant to which they are being sent; such acceptance must be officially recorded in an 'acceptance docket'. The producer must provide the haulier with information on the characteristics and dangers posed by the wastes to be transported, together with details of measures to be taken in the event of an accident.

In the Autonomous Community of Madrid, waste producers, transporters and operators are obliged to complete an industrial waste declaration note for every collection. These procedures of waste acceptance will be established in the new enforcing Regulations.

The degree of implementation and enforcement of the transport regulations is not known. However, it is unlikely that general compliance is being achieved.

3.4.12 UK

The requirements of the Directive 78/319/EEC on the transport of hazardous waste are enacted in the UK by the Control of Pollution (Special Waste) Regulations, 1980. These specify the procedures required for the movement of certain hazardous waste, some aspects of which are not required by the Directive.

A minimum of three days notice before shipment takes place is required by the Waste Disposal Authority so that the site earmarked for disposal can be assessed as regards it suitability to receive the waste load. The assessment takes the form of ensuring that the site is licenced to receive the waste and there is capacity in the site for that waste.

The strictness of the procedure may be relaxed for the producers of regular waste consignments in order to reduce the administrative burden on the waste handler and the disposal authority.

It is noteworthy that though the consignment note procedure is in accordance with the EEC requirements, there is no requirement in UK law to keep records of the production and storage stages of toxic waste management.

Waste transporters are subject to a range of statutory controls that apply to all types of businesses and not just hazardous waste transport. Most of this legislation is concerned with the safe operation of vehicles regardless of the nature of the load. The Dangerous Substances (conveyance by Road in Road Tankers and Tank Containers) Regulations, 1981, effectively consolidated the majority of current UK legislation affecting movements by road. Surveys carried out by the Health and Safety Executive have, however, revealed that one third of tanker vehicles checked were in breach of regulations.

Operators of goods vehicles over 3.5 tonnes gross weight are also subject to the system of operator licencing established by the Transport Act 1968. Under this system, the general competence and reputation of goods vehicles operators are periodically assessed. Whilst convictions under environmental legislation can be taken into account by the licencing authority, there is no obligation on it to do so. The use of the vehicle operator licence system as a means of raising standards among waste transport contractors is not satisfactory since it does not apply to vehicles under 3.5 tonnes. In this respect, the goods vehicle licencing system is not an effective deterrent to the illegal dumping of wastes or to the practice of fly tipping.

In examining this issue, the Royal Commission on Environmental Pollution - in its 11th Report, has recommended that the existing licencing system should be supplemented by a system of registration for the operators for all vehicles carrying wastes, irrespective of vehicle weight.

The Commission envisages that a system of registration would improve the security of the waste stream by :-

- helping the waste producer select a competent contractor;
- enhancing standards in relation to the manner and means by which wastes are transported; and
- ensuring that adequate information about the identity and movements of waste is available to those which handle and dispose of it.

The period of registration would correspond with that of the licence and an application of renewal would provide the licencing authority with an opportunity for an assessment of the operator's record, performance, training and other factors to establish whether his registration should be renewed. Such recommendations are still under consideration.

3.4.13 Summary of Transport of Hazardous Wastes

The area of hazardous waste transport legislation shows some general degree of uniformity between Member States. This is mainly because all countries have compiled with Directive 78/319/EEC in national legislation and implemented requirements concerning the transport, labelling and packaging of wastes. However, some countries have yet to enforce this Directive and others are only operating the transport documentation system (prior authorisation) on an irregular basis.

Some Member States have established additional requirements to those of Directive 78/319/EEC. These include the need for compulsory insurance by the carrier, appropriate authorisation for the waste haulier and the use of licenced carriers by waste producers.

A new Directive concerning the supervision and control of transfrontier shipments of hazardous waste is intended to impose a notification system and to fix common conditions of transport and packaging of hazardous waste. The impetus for this Directive has arisen because international trade in hazardous waste has been rising throughout the 1980s and with widely varying definitions of hazardous waste operating within the EEC, numerous difficulties have been experienced due to the appropriate labelling, storage and handling of certain highly toxic waste consignments.

The Directive has yet to be implemented and certain details concerning waste definition problems and the notification procedure have yet to be finalised. It is envisaged that this will be overcome in 1988.

3.5 Evaluation of Environmental Protection Legislation and Practice

With the implementation of planning controls and the obligations and duties under EC and national legislation, e.g. authorisation of waste plants; setting up of competent bodies; the notification of transportation; the requirement to keep records etc, it would appear that an adequate framework for the control of hazardous waste in Europe is provided. However, it is apparent that even in some Member States where legislation and the competent bodies to implement it have been established for many years that this 'framework of control' does not ensure a satisfactory level of environmental and public protection.

For example, in the UK, the legislation appears comprehensive, the waste disposal authorities in most cases have the expertise to fulfill their duties and many of the large waste disposal companies co-operate with the authorities and conform to the legislative requirements. The planning and site authorisation stages are also carried out satisfactorily. However, the day-to-day control mechanisms do not ensure uniformity or good disposal practice throughout the UK. The waste disposal authorities are also unable to effectively prevent or deter the illegal disposal of hazardous waste and when such illegal acts are discovered the penalties are insufficient to deter repeated offences.

In Member States where uncontrolled dumping is prevalent, it is likely that the competent bodies are not functioning effectively. This may be for a variety of reasons, e.g. inadequate training and resources, lack of information etc. Similarly, the lack of coherent and practical hazardous waste management strategies in these Member States may also be attributed to the inadequacy, whether in terms of resources or expertise, of the competent bodies and the responsible government departments.

In the following sections these and other problems relating to the environmental legislation and its enforcement are discussed.

3.5.1 Lack of Experience

Through the introduction of the Framework and Toxic and Dangerous Waste Directives, a degree of legislative uniformity has been sought and Member States have responded, either by introducing laws for the first time or amending existing national legislation. In some instances the national laws directly follow the wording of the Directives.

For some Member States, the process of compliance with EC legislation involved only minor adjustments and amendments to existing national legislation. For example, in the UK compliance was already largely achieved through the Control of Pollution Act, 1974. Similarly, in Germany, many aspects of the Waste Directive were already enacted. In other Member States, such as Greece, Portugal and Italy, very few control mechanisms existed prior to the Waste Directives and in these countries EC legislation has resulted in a very significant expansion of legal controls concerning the management of hazardous waste.

Because each Member State had a different legislative base-line from which compliance was sought, it is expected that it will take longer to achieve effective full compliance and satisfactory standards of hazardous waste control in the EC, than was first envisaged.

3.5.2 Acceptance Criteria for Waste Treatment and Disposal

The treatment processes at most hazardous waste facilities usually operate within strict parameters, the equipment and process being appropriate to only a narrow selection of wastes. Therefore, the suitability of a particular waste for treatment is determined by the design capabilities of the plant itself. Similarly, the equipment also puts constraints on the volume of waste throughput at such plants and, therefore, the quantity and date of acceptance must be controlled by the site operators. For these reasons specialist treatment plants often carry out pre-delivery waste analysis, and then, if the waste is suitable for treatment, a specific day for delivery will be agreed. Clearly, without such procedures, storage capacity can be overstretched and wastes that are unsuitable for treatment will arrive on-site.

The acceptance of waste at landfill sites is less critical in the sense that capacity constraints are not usually problematic. However, certain hazardous wastes may require specific unloading, safety or disposal measures and some on-site preparation may be needed. In such cases it is preferable that an agreed delivery date is arranged. This also allows enforcement officers to be present on-site to inspect the load and if necessary for sampling to be carried out. It is noted that these procedures of acceptance are implemented at some landfill facilities but by no means uniformly throughout the EC.

In Member States where co-disposal is carried out, the logistical problems of sampling and analysing are enhanced. If authorities wish to ensure illegal disposal does not occur, it would be necessary for them to sample all waste loads entering the site. However, at some sites with very high disposal rates, effective sampling and checking of loads would be time consuming and requires considerable resources in terms of staff and equipment.

Within the UK, recommendations have been put forward regarding pre-disposal sampling at co-disposal sites, however, on a national scale, this practice has only been adopted by a few contractors.

In other Member States, e.g. the FRG and France, hazardous wastes can only be disposed of at a restricted number of tightly controlled sites, and all waste loads entering such sites are sampled to check their conformity with the transport documents. In the FRG, for example, either a detailed quantitative analysis or a rapid qualitative test is carried out before any waste is landfilled.

It is therefore recommended that to ensure a greater level of environmental protection, increased provision should be made in order that sampling and analysis of waste is carried out before disposal or treatment takes place. The need for such procedures is particularly important for landfilling operations, where opportunities for illegal disposal are most evident, i.e. disposal, and where the current levels of pre-disposal sampling and analysis are either inadequate or non-existent.

3.5.3 Authorisation of Hazardous Waste Treatment and Disposal Operations

Authorisation prior to the operation of facilities treating or disposing of hazardous waste is required in all Member States. In most instances, the authorisation is based on proof of the technical competence of the operators and the assurance that adequate environmental protection measures are being taken. Variations do exist, for example, in France the authorisation is conditional upon the submission of an environmental impacts assessment, while, in the UK, an operating authorisation is only issued if planning permission, which involves consultation with experts in hydro-geology and water regulation, has already been granted. In the Netherlands a licence to operate a hazardous wastes facility is only granted if certain "effectiveness" criteria are satisfied.

In other Member States specific conditions concerning authorisation must be made but as yet these have not been widely adopted. For example, in Luxembourg, the authorising body can make aftercare or rehabilitation of a site a condition of the authorisation. In Spain and Belgium, financial guarantees or civil liability insurance may be imposed as a condition of the authorisation.

The setting and compliance with authorisation conditions is one of the primary means of ensuring public and environmental protection. Typically such conditions will specify the nature of the waste that may be treated or disposed of, their quantity, methods of disposal, unloading procedures for incompatible waste, etc. This type of control mechanism, if consistently applied and enforced should lead to uniform standards throughout each Member State. For some categories of treatment, e.g. incineration, the conditions and standards of operation show a degree of uniformity. In the case of landfill, however, examples of poorly specified operating conditions have been observed, e.g. the nature of wastes for acceptance are not specified and where conditions have been specified by the authorities, they are not always complied with. This situation is particularly evident in the UK.

Though compliance with EEC legislation has largely been achieved, problems of implementation have arisen in some Member States, concerning the authorisation of hazardous waste plant and the issuing of operating conditions. For example, where detailed guidelines on the technical requirements for approval do not exist, regional authorities can have wide discretionary powers and it is possible that approval in one particular region may be denied in another.

The process of seeking approval may also be subject to considerable administrative delays. The situation has arisen in Belgium, where the slowness of the authorities to issue permits results in plants commencing operation before authorisation has been obtained. Similar problems are experienced in the FRG.

In other Member States, some sites are operating, which have never been assessed for their suitability to receive hazardous waste, often because they pre-date the legislation, e.g. Greece. Clearly, it is a matter of urgency that such sites and facilities are reviewed and, where necessary, the operating conditions of the authorisation are modified if further (and extremely costly) environmental damage is to be avoided.

3.5.4 Waste Disposal Plans

A requirement of the Directive 78/319/EEC is that competent authorities should produce waste disposal plans which must be kept up-to-date and be made public. These plans must be forwarded to the Commission so that regular comparisons can be made between Member States. Such plans must include information on :-

- The type and quantity of waste arising
- The methods of disposal
- Specialised treatment centres where necessary
- Suitable disposal.

The Directive does not make it clear whether the term 'plan' is to be understood as a general statement of intentions or as a detailed enforceable scheme. However, where plans have been submitted they are more in-line with the former interpretation.

Some countries, e.g. FRG and Denmark and regional authorities within other Member States have drawn up satisfactory plans in line with the Directive. However, in France there is no legal requirement for such plans to be produced and in other countries, the task of producing plans has been tackled with little urgency.

The delays associated with the production of these plans suggests that the information base and level of forward planning for the management of hazardous waste is largely inadequate in these regions. If such plans containing accurate information were available throughout the EC, detailed comparisons of national policies and practices on waste disposal, treatment and recycling could be made.

In some countries the benefits of a comprehensive waste disposal plan are evident. In Denmark and the Länder of the FRG, detailed waste management plans have been drawn up and implemented, matching capacity requirements with needs, specifying the organisational structure and objectives to be achieved. Recently the Government of Luxembourg has developed and implemented a national policy on waste management which specifies objectives and the methods by which they are to be achieved. It is noted that the main theme of this policy is towards the prevention, recycling and re-use of waste.

Thus it is recommended that greater urgency be applied to the development of long-term waste management strategies and programmes. These should consider ways of: reducing the distance over which hazardous waste are transported; ensuring appropriate treatment capacity is available; drawing up recycling programmes and, where appropriate, co-ordinating actions bewtween neighbouring disposal authorities.

3.5.5 Hazardous Waste Storage

Waste storage is a major area of concern in some Member States, particularly in those where competent bodies have no information on where wastes are arising and being stored. This problem arises because the Directive 78/319/EEC does not provide a definition of storage and some Member States have failed to define it in their national legislation. Thus, companies generating waste are not required to keep records or notify the competent bodies of the wastes contained on their premises (e.g. UK). In contrast, some States have legislation which commands manufacturers to dispose of their waste (within a given time period) i.e. they are not allowed to store it, while in others (e.g. the Netherlands) storage is well regulated. In some instances where storage is allowed regulations specify conditions on how the waste must be stored. Therefore, it is recommended that a common definition of storage should be adopted by Member States so that records may be kept and made available to the competent bodies.

Related to this issue is the storage or abandonment of hazardous substances (waste) on disused and vacated premises. Examples of this have involved the abandonment of large quantities of highly reactive substances for several years without appropriate storage conditions. In some Member States the competent bodies do not have access rights to such premises and thus action to remove and treat the waste only takes place after an incident (fire, contamination, etc) has occurred or the new owner investigates the site.

It is recommended that the regional competent authority have access rights to all disused industrial premises and all premises that are about to be vacated in order that appropriate measures can be taken before an accident or pollution incident occurs.

Member States should provide a definition of the term 'storage' in their national legislation and that conditions should be attached to such storage in order to ensure the safe containment of the wastes.

3.5.6 Recycling

The Directive 78/319/EEC requires that Member States encourage the use of waste reduction measures and the re-use of toxic waste. This aspect of the Directive has, as yet, had little practical effect on the arisings and disposal of hazardous waste in the EC. The lack of progress arises because the Directive does not specify any mandatory obligations or objectives in this area. That is, the encouragement of recycling is not a strong enough requirement, particularly where commercial factors are operating and dictating disposal practice.

At a national level, few Member States have attempted to define or implement a policy or programme on hazardous waste recycling. Broad statements of intent are contained in national legislation and waste disposal plans but methods of implementation and specific actions are generally absent.

In some Member States progress in recycling has been made. In Germany, the operation of waste exchanges by the Chamber of Commerce and Industry and the Association of Chemical Industries are contributing to this improvement. Thus the quantity of waste being externally (not in-house) recycled is likely to increase further as the cost of disposal rises in response to the Technical Directive on Waste (TA-Abfall). In Luxembourg, the Government has produced an overall policy on waste management which is based on the prevention, recycling and re-use of waste. The policy will be accompanied by information campaigns and financial measures - the objective being to organise the selective collection of wastes for recycling on a national scale.

In most Member States there is little investment in recycling plant because of cheap disposal costs and fluctuating prices (of disposal and recycled waste). It is recognised that some Member States provide subsidies (tax concessions) to encourage recycling, but as yet, these have not been widely adopted in the EC.

It is recommended that greater efforts be made at national level towards the organising of waste recycling schemes and the eventual elimination of certain substances from manufacturing processes. National and regional disposal plans should include a comprehensive programme on waste reduction and recycling, specifying aims and a time scale for their implementation.

3.5.7 Enforcement

The analysis of national legislation relevant to the management of hazardous waste in the EC shows some degree of variation between the Member States. However, it is evident that the national or regional legislation is not the main determining factor on the standards of hazardous waste disposal. It is the enactment and enforcement of the legislation which have most influence. In some Member States appropriate enforcement bodies do not exist or are handicapped by lack of resources, unclear boundaries of responsibility or restrictive powers (eg in Greece, Portugal, and to a certain extent the UK). It is also apparent that the same control structure and legislative backing can be in place nationally, but because the vigilance and practice of enforcement varies, substantial disparaties can still arise. Not only does this result in varying levels of environmental protection but such regional variations can also be exploited commercially by less environmentally concerned companies, which, in turn, encourages the transfrontier movement of wastes.

It is noted that national and regional variations occur because some Member States and Regional Authorities have yet to introduce enacting regulations. Some countries are in the process of remedying this situation. However, unless the necessary regulatory body and authorities are in place to ensure implementation, a substantial improvement in practice is unlikely in the short term, even with the enacting regulations.

It is recommended that :-

- the actions, role and effectiveness of enforcement bodies should be reviewed; and

- where possible, the competent bodies and other regulatory agencies should co-ordinate their actions to reduce the amount of duplication and overlap of enforcement duties.

3.5.8 Illegal Practice

Illegal deposition of hazardous waste by mixing or hiding it amongst domestic or inert industrial waste is known to be prevalent in some Member States but there are no estimates on an EC scale which indicate the magnitude of the problem. Similarly, the illegal dumping and/or transportation of hazardous waste is not uncommon, which again suggests that many such incidents pass undiscovered by the waste disposal authorities. For example, in France, it is estimates that 200,000 tonnes of hazardous waste are disposed of illegally each year. In the Netherlands, it is estimated that only 20% of hazardous waste is not notified prior to disposal. Although this does not indicate widespread illegal disposal, it does indicate a degree of uncertainty concerning the monitoring of hazardous waste. The status of monitoring and control conditions in other Member States suggests that throughout the EC over 3 million tonnes a year are disposed of illegally without control and with unknown environmental and public health consequences.

The ease with which hazardous waste can be disposed of illegally also indicates that, in many instances, the function of the competent bodies in Member States is essentially one of record keeping and supervision, i.e. in general they do not act as effective policing agents. However, it is recognised that the need for the 'policing of hazardous waste treatment and disposal' is not so important in Member States where there are no price incentives for industry to find the cheapest, but possibly less environmentally acceptable, disposal route. Similarly, where consortia have been formed to provide disposal facilities to member companies on a non-profit basis, or where waste is disposed/treated in-house the need for policing is not such an urgent requirement.

In Member States where a company's 'good' reputation is considered important, a degree of self-regulation is evident. Also, through the formation of waste disposal trade associations with their own 'codes of practice', some improvement in operation practice is discernable. However, where private companies dominate the operational aspects of hazardous waste management and price competitiveness is strong, the occurrence of illegal or unacceptable disposal practices is likely to continue. Thus, reliance on self-regulation by the waste disposal industry cannot be depended upon for the attainment of uniform high standards.

3.5.9 Penalties for Illegal Practice

Within the EC, the penalties for illegal or negligent practice vary both across the EC and regionally within Member States. In many instances the penalties imposed on contractors who illegally dispose of waste are not considered to be severe enough. For example, fines are often not much higher than the cost of disposal and, thus, they do not adequately deter illegal practice.

Similarly, in most instances, it is not possible for authorities to prohibit a contractor from operating even though that contractor may consistently dispose of waste illegally. Therefore it is recommended that :-

- the penalties for illegal disposal and negligent action in the handling, treatment or disposal of hazardous waste should be of appropriate magnitude to effectively deter their occurence; and

- contractors that repeatedly infringe waste regulations or 'codes of practice' should not be allowed to continue to operate.

3.5.10 Resources for Enforcement

The lack of adequate resources available to competent bodies and other regulatory agencies is, in some countries, regarded as the main factor responsible for poor levels of enforcement and the continuation of inadequate waste disposal practice. The shortfall in resources can be due to a lack of trained staff and/or of appropriate facilities and equipment. For instance, in some Waste Disposal Authority areas in the UK, the level of supervision of hazardous waste disposal operators is often inadequate and staffing levels are such that 'spot-checks' and 'waste investigations' are not carried out regularly and, thus, contractors who adopt 'poor' practices have little fear of detection or prosecution. However, some Member States (e.g. Netherlands) have a comprehensive framework of enforcement involving Inspectorates, Provincial Authorities, police departments etc.

In section 3.5.11 below on Record Keeping the value of possessing accurate and up-to-date information on waste arisings for planning and control purposes is discussed. However, it is evident that many competent bodies in the EC are unable to benefit from such information because of the lack of adequate staff or efficient data collection systems that are necessary to keep records.

In many instances, the availability of resources for enforcement purposes is determined by the level of funding from the responsible regional or central government authorities. Thus, the enforcement bodies themselves, though they may recognise their need for additional resources, have little control over their own levels of resource allocation.

It is also evident that the costs of providing additional resources are not insignificant and that on a wider scale the costs of implementing EC legislation in general is a major factor contributing to the slowness and reluctance to attain compliance. Therefore, because some Member States are either unable or unwilling to increase funding, it is important that the structure and role of the existing enforcement bodies should be examined in order that they may maximise their effectiveness using the same resource base.

3.5.11 Record Keeping

The data available on hazardous waste arisings and disposal within Member States similarly show a high degree of variation between Member States. In Denmark, for example, an established control network exists which all chemical wastes pass through enroute to the only disposal and treatment plant. Also because enterprises must notify the presence of oil and chemical wastes to the local council, the arisings can be accurately correlated with the quantities treated or disposed. In comparison to Denmark, other Member States are poorly informed about the quantities of waste arising and being disposed of, even though there is a duty for producers and contractors to keep such records.

The issue of adequate and up-to-date record keeping is considered to be very important, not only because it enables the competent bodies to remain informed of waste arising and thereby draw up waste management plans, but also because a detailed information base can allow the competent authorities to carry out retrospective investigations. In order to do such investigations it may be necessary to carry out a comprehensive survey of all industry, making records of materials throughput and the annual quantities of waste generated. There should also be an obligation on the part of waste producers to inform competent bodies of process or productivity changes.

In some Member States the current system of collating information of hazardous waste arisings, transport and disposal is laborious and inefficient. Computer based inputting and data storage systems may improve the efficiency of the record keeping process, while making the data readily available for analysis.

It is recommended that data collection systems in Member States should be reviewed and, where necessary, alternative methods of collating data should be implemented.

3.5.12 Training, Information and Guidelines

The provision of and level of training in the management of hazardous wastes in the EC is variable. In only a few countries is specific training given to those who are responsible for enforcing the legislation as waste disposal/treatment facilities. In Germany, for example, comprehensive courses are run for persons who wish to be 'agents in charge', thus ensuring that these important positions are held by trained and qualified personnel. In the Netherlands, persons responsible for enforcing legislation receive specific training.

There are numerous examples of large companies providing training for all members of staff, covering aspects such as 'good practice', accident procedures, legislative requirements, etc. However, in most cases, this training is only available in-house. Frequently, within industry, responsibility for handling and keeping records of hazardous waste arisings is not a management activity; rather it is delegated to operatives with little formal training in waste management. In all of these areas, there is cause for concern, in that hazardous wastes can be transported and disposed by poorly trained workers without basic educational skills. In many instances, they are also inadequately supervised. This has led to a number of examples of wastes being disposed of either to totally unacceptable facilities or handled on-site with minimal safety provision. In the UK, a number of individual private sector companies have adopted comprehensive training of workers. The training of enforcement officers is almost entirely 'on-the-job' with few special courses available to enhance their skills. There are no degree level courses in waste management.

Thus, it is recommended that the training needs covering all aspects of waste management should be reviewed and that, where necessary, training courses should be developed and made available to all workers in the industry.

The guidance available to competent bodies and waste disposal contractors concerning methods of transport, treatment and the disposal of waste are particularly valuable if consistent practices are to be established on a national basis. It is also evident that where regulations or guidelines on procedures and technical aspects of hazardous waste disposal are provided or imposed, that they should be reviewed periodically and if necessary amended. This will enable developments in disposal/handling techniques to be introduced and faults and loopholes of current practice, which may have not been originally foreseen, to be corrected or amended. In some Member States such a review process is long overdue.

An important aspect related to this issue is that the waste management industry in some Member States is aware that overall standards of disposal are not satisfactory and that as a consequence of this more stringent legislation and new national policies may be introduced. Because the industry in general would not welcome such changes and in order to pre-empt this move some companies have responded by deriving their own codes of practice. However, experience suggests that the existence of such codes of practice does not ensure they are adopted by all members of such associations. Reliance on the voluntary adoption of 'codes of practice' is not being followed in Germany, where the introduction of a Technical Directive on Waste (TA-Abfall) will seek to ensure the harmonisation of standards by enforcing codes of practice and technical solutions on the waste disposal industry.

4.0 CASE STUDIES

4.1 Introduction

This section of the report draws upon the national case studies of hazardous waste treatment and disposal facilities to highlight examples of 'good' practice and to identify poor practice concerning overall safety standards in the handling and monitoring of hazardous wastes and clean up of contaminated land.

4.1.1 Normal Practice

It is evident that though a wide variation in operational standards and provision for safety exists within Member States some countries have similar waste management characteristics. On the basis of these characteristics, three broad categories of normal practice have been recognised in which Member States have been grouped.

Group 1 : Very High Standards. The provision of safety and environmental protection at treatment and disposal facilities in Denmark, FRG and the Netherlands is of a high standard throughout. The administrative and operational aspects of hazardous waste management are based on a high level of forward planning, stringent waste monitoring and sampling procedures and the presence of high profile workers and environmental protection authorities. Worker training at all levels is given a high priority as is the provision of worker safety and first aid equipment.

The FRG in particular is moving towards technical solutions to reduce the opportunities for workers to come into contact with hazardous waste. They are also more actively seeking solutions to the overall reduction of hazardous waste arisings through recycling and manufacturing process changes. Furthermore, technical directives on waste will continue the standardisation of waste treatment methods 'state of the art' processes.

Group 2 : Good Operational Standards. This group, comprising the UK, Belgium, Ireland, France and Luxembourg, are characterised by examples of good standards, particularly at large incineration plant and for the in-house facilities of major manufacturing enterprises. However, nationally, sufficient incidents of poor or uncertain practice and illegal disposal occur to cause concern. The provision of safety equipment, levels of enforcement and training is typically of a high standard within the larger waste disposal companies. In some instances, the company image and track record on worker and environmental safety issues is seen as a commercial asset. At small landfill and physical/chemical treatment operations, the monitoring and sampling procedures and the safety and first aid provisions are not always satisfactory. The poor standards of small operators is partly due the lack of adequate training and awareness of the risks involved.

Group 3 : Variable Operational Standards. The countries of Greece, Portugal, Spain and Italy are characterised by the wide divergence of operational and safety standards. Good examples do exist within large multi-national companies, manufacturing consortia and at specialist treatment/disposal companies. However, uncontrolled tipping, shortage of hazardous waste disposal and treatment capacity and unsatisfactory provision for worker and environmental protection are widespread.

It is noted that these countries are in a transitional stage and that much of the legislation on waste is new and has not yet been effectively implemented. Problems of inadequate management strategies for hazardous waste and co-ordination between various regulatory agencies are also evident.

These three groupings of 'normal' practice are only intended to provide general summary of the operational situation across the EC. To further illustrate some of the practices within Member States a series of case studies are now presented.

4.2 Country Case Studies

4.2.1 Belgium

In Belgium, internal safety procedures at sites receiving hazardous wastes, generally take the form of the conditions imposed under the terms of operating and building permits.

There is only one hazardous waste treatment centre in Belgium. It carries out a physico-chemical neutralisation process in the treatment of wastes containing inorganic products (excluding cyanide). It treats some 16,200 tonnes of waste annually and employs 8 persons. The plant is represented here as a good example of internal safety procedures at a hazardous waste facility. The operating permit contains regulations on the plant as a whole (storage arrangements, construction, monitoring and operating procedures). As part of the approval the operator must make available to the official responsible, documents proving the payment of the premiums due in respect of the civil liability insurance policy. The building permit contains provisions relating to firefighting equipment, heating installations, electricity, lighting, emergency medical facilities and regular inspections.

Following regionalisation in Belgium, there is no competent body set up in Walloon for the monitoring and inspection of plants and there appears to be insufficient resources and a lack of political will to set such a body up. Thus, monitoring and inspection are rarely carried out.

Staff representation in internal monitoring of safety procedures is totally random since there is no necessity for a Health and Safety Committee when firms employ less than 50 workers.

Staff training generally takes place on the job with few special training courses. The safety procedures which are most difficult to enforce are the wearing of headgear, breathing apparatus and goggles. This is largely because such protective equipment is uncomfortable to wear and is considered to reduce the 'efficiency' at which workers operate.

The physical/chemical treatment plant discussed above makes provision for basic internal safety procedures, however, considerable improvement in the organisation, enforcement and monitoring of safety procedures could be made. For instance there is no indication of the need for, or standard of, first aid training or the practice and frequency of emergency drills.

In addition, although the requirements of the operating permits help ensure basic safety measures and procedures, in practice the slowness of the competent commune and provincial authorities in issuing operating permits means that operations sometimes start before permits are obtained. The situation is exacerbated by the absence of any objective monitoring and the inadequacy of penal sanctions.

4.2.2 Denmark

Kommunekemi is the only major plant in Denmark for the treatment of oil and chemical wastes. It consists of two incinerators, an oil treatment facility, an inorganic treatment plant and a controlled landfill. Since its opening, Kommunekemi has been subject to the Environmental Protection Agency's direct supervision. An employee of the Agency has been dedicated to exercise this supervision since 1983. The plant is located very close to a residential area and the majority of the complaints received by Kommunekemi relate to nuisance caused by odours. The presence of a supervisory officer has proved useful in rapidly indentifying and remedying these incidents.

All operational aspects of the Kommunekemi plant have been separately approved by the Environmental Protection Agency, e.g. drum emptying, incineration, storage, water treatment, environmental protection, etc.

There are guidelines concerning gaseous emissions from the incinerators and solid wastes from these are deposited on site, at a controlled landfill. Surface run off from the plant is channelled to a common collection system, via an oil separation unit after which it is stored in holding tanks before discharge into the sea. This run-off is sampled before discharge and, if specific concentration limits are exceeded, the outlet from the holding tanks is closed and remedial action is taken.

The safety regulations of the plant are set out in the Safety Instructions Manual, which all employees can study during working hours. Introductory courses on safety, follow-up courses and a personal safety folder, which is also given to outside contractors, are provided for all employees.

The Waste Acceptance at Kommunekemi involves strict monitoring and sampling procedures for waste. The waste analysis normally determines the oil, solvent, organics, pH, odour, nitrate, sulphur content and its calorific value, though a more wide ranging analysis of the waste may also be carried out when necessary.

It is noted that during the operation of the Kommunekemi site, accidents have taken place involving gaseous releases, fires, explosions, etc but no persons have ever been seriously injured. These accidents have been investigated by safety officers and, as a result, a number of changes in operational procedures and processes have been made. Overall, a high standard of safety provision and degree of compliance is achieved at Kommunemeki and management efforts are continuing in order that risks and accidents can be further reduced.

The safety organisation is structured as set out in the national law, with a number of safety groups represented. Safety audits, manuals, emergency signals and medical examinations are examples of measures implemented on site. The high standards of the internal safety procedures at Kommunekemi are partly attributable to the comparatively simple and well structured system of hazardous waste management in Denmark. Such a system enables the management of the plant to be fully informed, at all times, of the nature and quantity of wastes arising and, thus, they achieve a high level of forward planning.

4.2.3 France

The setting up and operation of a centre for the treatment of toxic and dangerous waste comes within the field of application of a prefectoral decree on classified establishments and the labour code. This decree lays down and defines the body of regulations applicable to an establishment. It is a comprehensive document defining general and specific instructions on operation conditions, installation rules, structural regulations, environmental protection measures, waste acceptance and emergency procedures.

The Homburg physico-chemical treatment centre treats effluents mainly produced by metal plating works. There are numerous safety procedures which have been instituted and applied in the operation of the centre relating mainly to protective clothing, safety equipment, emergency procedures, equipment safeguards, transfer of liquids, safety data sheets, first aid training, etc. The centre comprises a laboratory, liquid and solid waste storage plant, a workshop and a line for the regeneration of ion exchange units. It receives waste from a large number of metal plating workshops within the region where wastes must be accurately identified before treatment. The centre has gradually acquired and developed its safety control procedures and equipment such that the internal safety procedures are now comprehensive aiming to minimise accident occurrence and effectively plan for emergency situations.

High operational and internal safety standards are also found at many other waste treatment and specialist incineration plant in France. For example, the Saint Vulbas PCB incinerator is renown for its high safety standards.

The plant has a small work force of 60, involved in the incinerations of liquid, sludge and solid industrial wastes, in particular, stable organics containing varying concentrations of halogenated compounds. The plants' capacity is 22,000 tonnes per annum, shortly to be increased to 27,000 tonnes following the commissioning of a new incineration unit.

Organic halogenated wastes are delivered to St Valbas in steel drums, (typically 200 litres). For operational purposes, the liquids are classified as high calorific value (HCV) and low calorific value (LCV) wastes and stored and treated separately. On arrival, the products undergo tests based on a special acceptance procedure to determine :-

i) whether or not each waste load conforms to the initial sample provided by the producer and whether it should be accepted for treatment or be rejected; and

ii) its treatment path (up to the final incineration stage), based on its physical and chemical properties (dilution, mixing, etc).

Thus, before any product is accepted, the producer is required :-

i) to fill in a standard identification form; and

ii) to forward a representative sample of the waste to the treatment centre.

This sample undergoes analyses for the purpose of identification and verification to supplement the information originally supplied. In addition, test treatment is carried out in a laboratory simulating the proposed industrial treatment, in order to establish the efficiency of the treatment method and the nature of the residual wastes.

Specific components of the plant include: a rotary furnace capable of handling a wide range of waste products; a LCV liquids store; a HCV liquids store; a drum storage area; a bulk storage area for pulps; and a pit for bulk solids. Emergency equipment facilities used at the plant includes foam generators, powder tanks, fire hydrants and portable extinguishers with powdered carbon dioxide or atomised water.

Safety assessments are carried out regularly which examine the causes of accidents, e.g. fire, spillage, and other process risks related to liquid storage, dumping areas, furnaces and the treatment and recovery workshops.

4.2.4 Germany

Incineration in Germany is regarded as one of the most environmentally favourable methods of treating special wastes and numerous publicly, public/privately owned and private incineration plants are operational. Such plant utilise state of the art technology, particularly in view of the statutory obligation to apply specific techniques to protect the environment (e.g. comply with very strigent emission standards as laid down in the TA-Luft) and to review the technology currently being used on a regular basis. Incineration is currently used for approximately 8% of hazardous wastes.

The waste acceptance procedures of each plant will depend on the nature of the waste received. However, automatic handling equipment is used in many cases which has both efficiency and worker safety benefits. Secondary combustion chambers ensure highly polluting wastes (solvents, paints, etc) are destroyed. Temperatures of between 1000 and 1100°C are achieved during secondary combustion. Waste gases from the process are cooled and first passed through an electronic precipitator and then ducted to a wet scrubber where chlorine, fluorine and sulphur dioxide are removed. The clean gas is reheated before venting through a stack.

The whole combustion process is co-ordinated from a central control room. Sensitivity safety chains are interlocked in such a manner to ensure safe operation. Flue gases are monitored continuously and emission measurements in the area of the plant are conducted by both the Technical Supervisory Services (TUV) and the Bavarian Office of the Environment (respectively the responsible authorities of other Länder). The in-house laboratory has also continuous monitoring programme, sampling wastes as they arrive at the plant and all safety devices, instrumentation and controls are subject to constant checks by TUV.

Operating permits are required by all plant. Its approval is conditional on strict safety and operating procedures being met. Personnel are highly qualified and receive intensive training in the operation of the plants and first aiders are present on site at all times.

It still remains, however, that some 31% of special wastes are disposed by landfill. The practice of tipping special waste with domestic waste is no longer carried out. Current landfill practice requires that sites are effectively sealed, preferably with clay. Seapage at sites is captured by base filters, purified by effluent treatment plant, and then introduced into the main outfall. Landfills for hazardous waste should not be made secure from leakage only by artificial sealing. Where necessary, bentenite sealing of the sides is carried out.

Waste arriving at such sites are checked on delivery. Depending on the nature of the waste they will be deposited in different and separate disposal cells. The acceptance test will involve either :-

- A visual assessment, pH test and rapid analysis (5-10 minutes)
- Operational or in-depth analysis requiring extensive laboratory facilities

Qualified and experienced personnel are responsible for these tests and the space allocation for the waste. The rapid analysis is to determine the qualitative proportions of certain substances (chromate, cyanide etc) and thus to decide in, general terms, whether the waste is suitable for disposal. Qualitative analysis may also be conducted.

For the clean up of contaminated sites detailed worker safety rules are established on the basis of an action plan. These rules are compiled into instruction sheets which are continuously up-dated in line with in-coming information from analytical findings. The action plan and instruction sheets are concerned with the safety of personnel and visitors. Comprehensive safety measures are specified for activities and for the use of equipment, for example information concerning body and respiratory protection is specified for: drilling, handling, excavation and machinery handling activities.

Other provisions concerning mechanical check-ups, procedures for site visitors are also specified. In particular, because of the high risks of unexpected exposure to dangerous and/or carcinogenic substances, medical checks are required for all personnel working on site.

All the facilities described above and others in the FRG, that involve the handling, treatment and disposal for hazardous waste, the overall provision for worker and environmental safety is of a high standard. The compilation of worker safety action programmes, the policy of monitoring and sampling of wastes are considered to be particularly important.

4.2.5 Greece

To illustrate internal safety procedures in Greece two examples of hazardous waste disposal are discussed. One example concerns the Ano Liosia landfill, which for Greece, is representative of good practice. It cannot therefore be taken as an example of how other smaller, more numerous and uncontrolled sites are operated, which are still subject of much concern by the regulatory authorities.

The Ano Liosia site is operated by the Local Authorities. It employs 65 members of staff. Basic precautions, aside from methane venting, include a rudimentary separation of wastes (eg "household", organics, acids) and control over the burying of drums containing toxic waste. Because a large number of employees come into regular contact with the waste such employees are encouraged to undergo medical screening every three months. In the case of illness employees are carefully screened for clinical effects due to the handling of wastes.

The provision of first aid facilities is, at best, satisfactory. There are basic medical supplies but no accident record book is kept. More importantly there is no provision for eye baths, showers or 'dosing' facilities in the case of spillages or accidents involving liquid or gaseous wastes.

The inspection and monitoring of compliance with operating and safety conditions is irregular and thus the enforcement powers provided by the legislation are not currently fully exercised.

The other case study example is the operation of an oil refinery with on-site waste treatment facilities. Until recently the refinery was owned and operated by a major oil company and hence the current health and safety procedures have been largely inherited from this company. Accordingly, health and safety provision at the refinery are of a reasonably high standard. Protective clothing and respiratory equipment are provided to employees and there is a general good appreciation amongst employees of safe working procedures, particularly in the handling of toxic sludges. A very high standard of first aid facilities is also available on the site.

The above are examples of good practice and it is known that many other waste disposal and treatment operations in Greece have a much lower regard for worker safety with minimal provision for accident and emergency situations. However, the oil refinery example does illustrate the practice and policy of most large multi-nationals in deriving and implementing comprehensive internal safety measures to standards that are internationally uniform. The high standards of the Ano Liosia landfill site in comparison to other landfills is due to a number of factors :-

- The site is operated by the Local Authority

- Its management is supervised by the Association of Local Authorities

- The site accepts large quantities of wastes and thus its efficient operation requires a high level of formal organisation and management involvement.

4.2.6 Ireland

In consideration of internal safety procedures in Ireland an incinerator, chemical treatment plant and solvent recovery plant are briefly discussed.

The incinerator accepts a range of mixed organic wastes, solvents and process gases achieving a destruction efficiency of 99.9% with 90% SO2 removal by a caustic scrubber. A range of process controls are built into the incinerator to ensure operating parameters are maintained. Constant monitoring of exhaust gases as well as environmental monitoring are carried out. Appropriate protective equipment and clothing are used, including gloves, goggles, helmets and breathing apparatus as required. Regular fire drills and emergency exercises are carried out. All process control staff have a high level of expertise.

The physico-chemical treatment plant is at a developmental stage, but is still required to observe the operating conditions set down by the local authority in the planning permission and waste permit. The detailed conditions set out in the permit require identification of the various storage and process areas in the plant, full documentation of all operations, pre-operational plan, appropriate bunding and segregation of wastes, details of accident prevention, fire fighting and emergency response measures. The permit also requires compliance with the appropriate EC directives on worker and environmental protection and public safety. Other important aspects include analysis of waste prior to acceptance, proper labelling of packages and requirements concerning staff qualifications and training. The local authority is carefully monitoring the operations of this facility and ensuring the safety of the environment and the public. The Industrial Inspectorate regularly inspect the premises to ensure safe working conditions.

The environmental standards to be met in the solvent recovery operations are contained within the original planning permission. No permit has been issued under the hazardous waste regulations despite the apparent requirement for such a permit in national law. In this case the local authority does not consider a waste permit to be necessary for solvent recovery, since it does not regard recycled solvents as waste.

In plant of this type the production personnel and environmental management personnel are closely integrated and this linkage helps ensure that high standards of safety, training and personnel protection are applied. Both the Industrial Inspectorate and the local authority officers responsible for environmental control regularly inspect the plant.

The case studies of hazardous waste facilities and operations in Ireland illustrate the advantages of having a responsible and effective control authority that seeks to implement operating conditions imposed by the planning and permit system. These factors are particularly important where company policy on internal safety procedures may be inadequate.

4.2.7 Italy

In Italy some companies operate their own Environmental Protection and Safety Department which is responsible for regulating and controlling the movement and correct disposal of waste. A procedural manual exists for the unloading of special wastes generated by companies which have disposal sites attached to the production facility. This specifies in detail the labelling, documentation, authorisation, first aid, monitoring, classification requirements and general responsibilities of the plant management, process workers and the Environmental Protection and Safety Department. There are also criteria concerning the management of all activities connected with the treatment of toxic waste by outside companies. In general, a high level of competence and operational safety is achieved.

Another case study concerns the thermal destruction of liquid wastes generated by a company manufacturing fine chemicals and pharmaceuticals. The plant consists of a combustion chamber, venturi scrubbers, cleansing tower and cartridge filters. Controls are carried out in accordance with regional regulations and continuous checks are prescribed for the post combustion chamber and chimney. Micro-pollutant checks are carried out by the local health unit. The established safety procedures refer to start-up operations and emergency measures. Manuals for internal use for specific dangerous activities are currently being prepared. Efforts with respect to improving staff training are also in progress.

The improvements and internal safety procedures and environmental protection measures taking place at the plant arise from the pressure of local opposition, the site being located near a built-up area.

The stimulus to change the operation of the plant in a major way has led to the restructuring of production cycles, and the adoption of new processes which are generally safer from an environmental point of view.

Some waste disposal facilities in Italy are owned and operated by consortia of companies which generate waste. All companies in the consortium dispose of their waste at cost price. The internal safety procedures at such plant, whether incinerators or landfill sites, are of a high standard. For example, vehicles entering the site with liquid waste are checked with a 2 metre pipette in order to investigate stratifications of the liquid. Analysis is carried out to make sure it corresponds to the data on the original contract. The incinerators are monitored and the data are forwarded to the local health units. Workers receive medical checks and have the use of face masks with filters, breathing apparatus and other protective clothing. The operational policy at the plant is to contain waste in sealed circuits, thereby minimising the exposure to workers. The company does not issue internal regulations on the safety of personnel, because it is considered that the containment policy (liquid waste is moved by means of pumps and pipes) largely obviates such a need. At a similarly organised landfill site (consortium ownership, non-profit basis) vehicle checks are carried out on each load entering the site. Analytical checks on the water table and leachate are made every 2 months. However, it should be noted that in dry conditions, which are not uncommon, dust and the toxic substances within it may contaminate personnel and the surrounding environment.

Overall the case studies in Italy indicate that the provision for worker and environmental safety is variable depending on the nature of the operation, the company structure and its policies. Standards are usually good at plant that specialise in treating/destroying large quantities of potentially dangerous wastes. This is largely because the parameters for treatment are such that the exact composition and quantity of the waste need to be accurately known. Thus, such plants operate in a similar way to a manufacturing plant, ie. because the treatment process is often complex, efficient organisation and management structures are required. The expertise of the staff at such plants, their awareness of the potential dangers, and the responsibilities of the safety and environmental control bodies all lead to the development and implementation of comprehensive safety procedures.

In contrast, landfill is a comparatively low technology operation, with few trained staff involved a relatively simple operational structure. The magnitude of the and perceived health risks are also usually smaller and the provision of internal safety procedures usually reflects this. In some instances protection of the environment appears to have a higher priority that personnel safety.

4.2.8 Luxembourg

In Luxembourg, the internal safety proceeding are established by the conditions imposed under the terms of the operating permits. Officially, there is no plant specifically for waste treatment, however, an incinerator and a waste dump are currently operational. The incinerator is a classified establishment and the authorisation for the plant lays down emission standards aimed at reducing possible harmful effects on the environment.

Little is known about the safety measures carried out at the plant or the degree of enforcement and compliance.

4.2.9 Netherlands

Within the Netherlands the Nuisance Act and Chemical Waste Act contain various requirements relating to the working conditions at hazardous waste facilities. However, the principal legislation used to regulate working conditions is the Working Conditions Act. For example, during the treatment at contaminated sites a company must :-

- Determine the composition of the contamination before the soil is unloaded
- Ensure that odour and dust are minimised
- Cover contaminated soil at all times
- The site is well maintained.

In this case study, general policy on safety is formulated at management level and supervision of safety procedure is carried out by a safety expert. Many aspects including medical surveillance, accident reporting, training accident procedures and personal protection are specified within the safety policy.

At a chemical waste incineration plant the policy is aimed mainly at the application of safe working methods, the prevention of contact with hazardous substances and effective measures to minimise the occurrence of accidents. The companies policy is made clear in a 'Safety Policy' memorandum. Senior management, safety experts, the company doctor, emergency services and employees are all involved in co-ordinating, modifying and implementing the safety policy. Training on the nature of the risks, work duties, health and safety and accident procedures is given to all employees. Though overall, provision for safety appears adequate, a number of minor problems remain. For example :-

- waste materials are sometimes found in damaged packages and must be repacked;

- stored drums are sometimes not labelled and storage arrangements do not allow adequate separation;

- extractors are not well located;

- containment of vapours and fly ash is not adequate.

From these examples and the other case studies conducted at hazardous waste facilities in the Netherlands, it appears that the establishment of good working conditions was largely based on compliance with the Working Condition Act. The employer's obligation to instruction and training, the reporting of accidents and the appointment of a safety expert are being fulfilled and are mostly of a high standard. Some employees, however, are still insufficiently aware of the dangerous aspects attached to the treatment of chemical waste. This situation may lead to poor compliance with the internal safety regulations. Thus the need for strict supervision by safety officers remains important.

4.2.10 Portugal

Case studies are presented for Portugal on incineration, and physico-chemical treatment and recovery plant. It should be borne in mind, however, that uncontrolled dumping is the predominant disposal method in Portugal. Internal safety at such sites has a very low priority.

The general safety measures in operation at an incineration plant for pesticide containing wastes, were generally good. However, the incinerator does not have a gas-scrubbing system and it is not known whether the flue gases are monitored. Incinerator workers receive instructions concerning correct operation of the plant and have to wear protective clothing, masks etc. Blood tests are carried out every month on personnel in close contact with wastes and regular safety information sessions are carried out. A safety committee is also operated consisting of management and worker representatives. The most important factor here, in relation to internal safety measures, is that the enterprise follows the general pattern of safety measures enforced by its German owned parent company. These measures are very strict and ensure the wastes are handled in a safe way at all times. Even so, environmental concerns still receive low priority.

A similar standard of provision for internal safety is practised at a plant for the incineration of a range of organic solvents and isocyanates. The incinerator has a caustic soda gas-scrubbing system and operates between 1000°C and 2000°C. It is not known whether the flue gases are monitored. The safety measures include the supply and use of protective clothing, breathing apparatus, activated carbon emergency masks and fire extinguishers. The plant has a medical programme which ensures that workers receive regular medical examinations including electrocardiograms, blood texts, x-rays, and ear and eye tests. All technical operating procedures take full account of safety aspects.

Internal safety measures at a plant receiving liquid effluents (acids, alkalis, compounds of tin, lead and zinc) are also comprehensive. In addition to the measures outlined above, monthly environmental tests are conducted in all risk areas and absorbent and neutralising materials are strategically placed for use in case of accidents and spillages and, thus, a high profile, safety policy is maintained in order to increase worker awareness.

Again, the provision for safety at these specialist plants is generally good. However, there is some concern about the adequacy of the environmental monitoring. Also, in practice, the level of worker compliance with the measures is also unknown.

4.2.11 Spain

The Spanish case studies discuss procedures at a physico-chemical treatment plant and a disposal site in the Autonomous Community of Madrid. The licences of both plants lay down general operating conditions, compliance with which is monitored by the Directorate-General for Town and County Planning, the Environment and Housing. Weekly inspections are carried out by the Directorate's analysts, weekly meetings are also held at the Directorate-General.

Safety measures at the treatment plant include pre-treatment analysis, security precautions, danger area notices, emergency showers, protective equipment and fire extinguishers. The plant has its own drainage containment system and the underside of tanks are examined daily for leaks. All other critical equipment (pumps, pipes, valves etc) are checked regularly. Emergency procedure plans have been drawn up which specify chains of command, actions to be taken, requests for assistance, plant shut-down and evacuation for when human life is in danger.

At the disposal site standard safety procedures are specified for the laboratory work carried out during the testing of waste loads. On-site the supervisor is responsible for holding monthly safety reviews, demonstrations of safety equipment, and ensuring protective clothing is worn. New employees are informed of all safety regulations. An emergency and evacuation plan has also been compiled in line with that of the treatment plant.

These two operations have received considerable attention from the Directorate-General of Town Planning, the Environment and Housing and this close supervision has enabled the development and implementation of satisfactory safety provisions. However, little is known about the general level of safety at other plants throughout Spain. The two cases discussed above should not be considered as representative but rather as examples of standards that can be achieved when the appropriate commitment and supervision are applied.

4.2.12 UK

Case Studies have been carried out on incineration and chemical treatment plant, land decontamination and co-disposal operations.

At the incineration plant, where a variety of toxic and intractable waste are treated, health and safety procedures are given a very high priority. Training in emergency procedures and general safety operations is given to all employees and regular lectures, seminars and drills are carried out to enforce these procedures. Multi-stage verification and monitoring of the waste is carried out to ensure waste entering and leaving the site does so safely and that the area surrounding the plant is not contaminated.

Other aspects of the internal procedures include :-

- Induction training
- First aiders always on site
- Staff whose main responsibility is to enforce safety
- Failsafe systems on the incinerator
- Health monitoring of employees.

It is considered that though the precise safety procedures at specialist incineration plant within the UK do vary, they are generally of a high standard. Because of the potential risks of widespread environmental contamination, both the public and statutory authorities demand very high operational standards for plants incinerating hazardous waste. Therefore, in comparison, some sectors of the waste disposal industry, operational safety procedures are both more stringent and effectively enforced.

The land decontamination case study also revealed a careful and professional approach to the issues of operational safety. Training schemes, protective clothing and provision for emergency situations were similar to those of the incinerator. The company only uses waste transport contractors with 'good' track records and requires that the most appropriate transport route is taken (minimum potential risk) by such contractors.

Another major aspect of safety during decontamination procedures is the need to plan for a flexible safety response, because the risks to personnel remain largely unknown until the site investigation has been completed. In this respect the training and experience of site personnel are invaluable. Employees with knowledge of chemical properties, their toxicities , mobility and reactivity are in practice, more safety conscious than persons with little practical experience.

The case studies of landfill sites licenced to receive hazardous waste indicated that operational safety procedures varied considerably. At some sites, worker safety is given a higher priority than profitability and a strict company safety policy is enforced. Other sites failed to implement even basic health and safety codes of practice, being run with a minimum provision for worker health and safety or good waste management practice. Problems of effective enforcement and worker compliance are perhaps more difficult to achieve at landfill facilities than at specialist treatment/disposal plant because of the lower level of supervision and the site operatives themselves are frequently reluctant to adopt safety measures which can be uncomfortable and inconvenient in terms of operational efficiency.

4.2.13 Summary and Conclusions from Case Studies

The following general points have emerged from the national case studies :-

- The safety procedures of large companies (including multi-nationals) appear to be more comprehensive and more effectively enforced than at smaller companies.

- Larger operational facilities usually have safety committees with worker and management representation. This increases the awareness, involvement and self responsibility of all workers.

- Where company policy on safety is inadequate, a vigilant and effective control authority is necessary to maintain basic levels of safety provision.

- Specialist treatment/disposal operations which handle large quantities of dangerous and toxic wastes operate under strict and formalised procedures. In such cases efficient operation can only be achieved through high standards of general 'housekeeping' and the minimisation of accidents and emergency situations. These factors and the high standards that are required by control authorities ensure such plants implement comprehensive safety and environmental protection measures.

- Examples of satisfactory or good practice are found in all Member States, but these are often not typical of the country as a whole. There is still some uncertainty of 'normal' operational practice.

- There are opportunities within all Member States for those companies with good internal safety procedures to communicate their policies and ideas on safety to other enterprises in the waste management industry. An official or formalised scheme with these objectives may assist this process and lead to a greater standardisation of safety throughout the EC.

5.0 CONCLUSIONS AND RECOMMENDATIONS

5.1 Introduction

The problem of hazardous waste is an emotive subject and increasingly regulatory agencies and the public are becoming concerned at the way in which such wastes are handled and disposed of in the EC. The principal aim of this work is to review current legislative provision and practice associated with the handling and monitoring of hazardous wastes and to provide guidance to the European Community as to the adequacy and safety of the hazardous waste handling and monitoring.

There are three main objectives to the current study. These are :-

- to provide a comprehensive picture and comparative analysis of the current legislation and other controls relating to the safe handling and monitoring of hazardous waste in Member States of the European Community;

- to indicate areas of improvement to, or appropriate alternatives to existing measures and practices by identifying those procedures which have been successfully implemented in some Member States; and

- to provide information to Community institutions which will assist in the development of appropriate waste management policies in the future, as well as indicating priority areas of research and/or further study.

The study has addressed the first of these objectives by summarising, in Sections 2.0 and 3.0, existing legislative requirements and in particular its implementation and enforcement within Member States as they relate to health and safety at the work place and environmental protection respectively. Having presented the legislative summaries for each Member State, examples of good and bad practice are taken from the case studies in Section 4.0 to demonstrate the extent to which common practice at hazardous waste facilities departs from the requirements set out in the legislation.

On the basis of this comparison, unsatisfactory practices have been identified and are attributed either to gaps in legislative provisions or some other failure or inadequacy associated, for example, with training, advice and information needs, methods of enforcement and/or resource constraints etc. Once the source of the problem has been identified, it is possible to indicate ways of improving or modifying hazardous waste management procedures (as required by objective two of the study) and to provide useful additional information to the Community for policy development purposes and in order to identify research needs and priorities. This information has been presented in the form of a series of conclusions and recommendations in Sections 2.0 or 3.0, the principal elements of which are summarised below.

5.2 Findings of the Study

5.2.1 Definition of Hazardous Waste

There are clear obstacles to the implementation of high standards of hazardous waste management within Member States arising from the various different interpretations of the term 'hazardous waste'. Problems of definition not only affect the practical management of hazardous waste on a day-to-day basis, but they also reduce the utility of information provided on waste arisings and, more importantly, inhibit the long term planning for wastes, which is a fundamental requirement of European waste legislation. Thus, the major difficulty is that definitions are not uniform throughout the Community, which allows for certain wastes to escape the rigours of the control procedures set out in the legislation. If a waste is defined as a hazardous waste then the cost of its disposal can increase significantly, which in turn encourages transfrontier movement of such wastes between Member States. There are additional administrative requirements (e.g. consignment note), which must also be attended to when wastes are defined as hazardous. These requirements and those concerned with monitoring, analysis, safety provision etc can all vary from country to country because of national differences in the definition and its interpretation of hazardous waste.

The proposal for a Council Directive on hazardous waste is welcomed and should provide a better definition of hazardous waste. It will also allow for the more effective implementation of the Directive on the transfrontier shipments of hazardous waste.

5.2.2 Legislation on Health and Safety at Work and Practice

The general legislation within the various labour and/or health and safety regulations of EC Member States appears, with some exceptions, to be adequate for the protection of workers employed in the waste industry. However, there are gaps in the legislation which are of significance in some Member States. For example, in Ireland the limited scope of existing worker protection laws in relation to premises covered, means that up to 80% of the entire workforce is not covered by health and safety provisions. Similarly, there are doubts that adequate health and safety provision is afforded to workers in contact with hazardous waste in Italy.

A second important gap relates to medical surveillance and this affects the majority of Member States, with the exception of Denmark, Belgium and the Netherlands, where legal requirements for health monitoring exist. Long-term health monitoring of workers in contact with hazardous wastes is crucial in that it provides the only means of assessing the effectiveness of preventive measures and long term occupational health risks associated with the hazardous waste industry. It is, however, recognised that in a number of Member States the type of worker involved in waste disposal is more transient creating difficulties for adequate monitoring. Furthermore, current assumptions about the adequacy of existing health and safety provisions are made in the absence of good statistical data on health aspects. These clear deficiencies in European health and safety laws are expected to be overcome by future amendments to the legislation, largely prompted by the recently formulated EEC legislation on occupational hygiene.

Although, in general, sufficient legislative protection for workers dealing with hazardous wastes exists, in practice poor standards of health and safety provision operate at many hazardous waste facilities in Europe. Examples include very basic health and safety provision at landfill sites in Greece, Spain and Portugal and variable standards of provision in the UK and Belgium.

At the disposal company level, problems appear to be most significant for small firms with limited resources and for enterprises which do not have a clear company policy on safety issues. The difficulties of these companies are compounded by a general lack of information and training on safety matters on the part of both the employer and employee. For example, a lack of knowledge and training in safety issues has been identified by the Netherlands as being the cause of at least some bad practices and subsequent health problems at hazardous waste disposal plants.

Other difficulties affect the regulatory arm of the hazardous waste industry, whereby limited resources, a lack of inspection staff and poor quality of training are obstacles to the efficient enforcement of health and safety regulations. Thus, in the UK where staff resources are under pressure, a heavy reliance is placed on the ability of inspection staff to accurately allocate limited inspection time to only the most serious infringements of health and safety regulations.

Difficulties also exist at the national level with at least one Member State, Belgium, reporting problems with enforcement as a consequence of the differing responsibilities of national and regional administrative bodies. Other Member States, particularly those with a relatively limited history of industrial health and safety provision such as Greece, are constrained by a lack of experience in implementing and enforcing health and safety legislation.

5.2.3 Environmental Legislation and Practice

The EC Directives on waste provide a legislative framework that is intended to ensure public and environmental protection and control over hazardous wastes from the cradle to the grave. The national legislation of Member States tends to follow a similar line in that a framework for control has been established. A framework approach to control, however, is not always adequate as evidenced by widely varying standards of waste disposal that exist throughout the EC from good standards and strict enforcement to widespread illegal disposal in unlicenced and often totally unsuitable landfills with no enforcement. It may be argued that it is not the legislation that is at fault but the enforcement by the Member States. However, in many respects the duties and requirements of both the EC Directives and national legislation are ambiguous in the sense that the 'definitions of effective compliance' are not established.

Ambiguities of compliance should not be problematic where regional or national guidelines are available or technical directives have been enforced, but in some Member States the guidelines have not been enforced by industry or by the competent bodies.

Other factors also impede the attainment of effective enforcement and compliance. These include a lack of appropriate guidance on good practice for disposal of certain wastes, the provision of inadequate resources (staff and equipment), the lack of or inadequate training of enforcement staff and the inadequacy of record keeping systems in providing accurate up-to-date information.

Illegal disposal of hazardous waste continues to be a problem in some Member States, primarily through uncontrolled tipping and the deliberate mixing to dilute the waste (e.g. Greece, Italy and Portugal) and hiding of waste.

The penalties for illegal disposal are clearly inadequate in some Member States, where the fines are typically of the same order of magnitude as the costs of disposal. Thus the offender, if successfully prosecuted, may not be sufficiently discouraged from this illegal practice.

In many instances, the competent bodies only provide a supervisory and record keeping function and they have few powers to control the flow of wastes or effectively police and investigate the incidence of illegal disposal.

In some Member States it is evident that the long distance transportation of hazardous waste is encouraged by the existence of regional price differentials for disposal, poor enforcement and lack of appropriate infrastructure (e.g centralised treatment facilities). This enables contractors to dispose of waste contrary to recommended codes of practice. Many competent bodies are aware of this situation but, because of their limited powers to direct the flow of waste, they are unable to prevent it. This problem is enhanced in some Member States because the regions over which competent bodies have jurisdiction are often very small. Thus, although they can control the level of enforcement in their own area, they have no powers over the procedures taking place in neighbouring regions.

In many Member States, disposal plans have yet to be compiled by the competent bodies. A shortage of information concerning arisings of wastes and of suitable treatment/disposal capacity are clearly causing problems in the drawing-up of such plans. Additionally, the term 'disposal plan' itself is an inappropriate term because, in many instances, competent bodies only have a minimal influence over how and where hazardous waste are disposed, as in the case of the UK.

The disposal of hazardous waste by co-disposal (the disposal of toxic chemical wastes with domestic refuse) continues to be an area of concern. This is largely because the standard of waste monitoring and sampling of such sites is often inadequate. At containment sites designed specifically for the receipt of hazardous waste, e.g. in France and Germany, all wastes are monitored, and yet in other countries where co-disposal is carried out and, perhaps the need for monitoring is even more important, the monitoring is either infrequent or non-existent. Co-disposal remains a cheap disposal route which, as already stated, is often inadequately supervised and monitored. Under these circumstances, it is difficult to envisage how disposal authorities or contractors, who either carry out or supervise co-disposal, can maintain that such methods ensure environmental protection.

The Toxic and Dangerous Waste Directive has had little impact in the area of waste prevention, recycling and re-use. All Member States have, until recently, neglected this aspect of waste management. There is no evidence in the work to date that recycling of hazardous waste creates environmental and health problems, except perhaps in the case of the incineration of waste oils, where stringent controls are being introduced.

The standard of record keeping in the EC is variable and, in many cases, unsatisfactory. Of particular concern is the lack of up-to-date data on arisings and storage. Clearly, poor data quality also undermines the value of the waste disposal plans drawn up by the competent bodies.

The provision of and level of training for the management of hazardous waste in the EC is variable. In only a few countries is specific training given to those who are responsible for enforcing the legislation as well as workers who operate waste disposal/treatment facilities. This is an area of concern in that hazardous wastes can frequently are transported and disposed of by poorly trained workers without basic educational skills.

5.2.4 Case Study Findings

The provision for worker and environmental protection and the degree of compliance at in-house treatment/disposal facilities attached to large manufacturing plant are generally of a high standard and the health and safety practices are often an extension of those used during the production stages. At large and physical/chemical treatment plant the overall provision of worker and public protection is satisfactory. At such plant, the safety procedures are often integrated with the treatment and process stages, i.e. technical process solutions are used to minimise accident risk and exposure dangers. In many companies, the provision, for and enforcement, of safety procedures has been a continuous process, as lessons learnt from past accidents are implemented and new equipment and treatment procedures are developed. Therefore, the experience in the handling, treatment and disposal of hazardous waste has been an important aspect in the development of good safety policies and practice.

In some Member States, the treatment and disposal of hazardous waste is an issue of serious public concern and the companies involved in the industry are under constant scrutiny. This has had the effect of making work and environmental safety a high priority area within the companies' overall policy objectives.

The provision for worker and environmental protection and compliance by the workforce with safety procedures is less satisfactory at landfilling operations where co-disposal and uncontrolled dumping are carried out. Supervision is poor and workers with responsibilities to enforce safety procedures often fail to fulfil their duties. Where low standards are observed they tend to reflect the low priority that management place on safety issues.

The inadequacy of safety provision at some landfill operations is partly due to the costs of such provision. Good safety procedures can also impede the efficiency, in production terms, of the landfill operation, thus indirectly raising the costs of disposal. For example, waste analysis, vehicle inspection and cleaning, safety drills etc, all add to the cost of the operation without increasing the waste throughout. In a competitive market, management are naturally reluctant to adopt practices which reduce the profitability of their operation. Some examples of strict and enforced safety practice at co-disposal sites have been observed and in such cases, the management have fully accepted the financial cost in order to maintain high levels of worker and environmental protection.

While it is recognised that many larger companies in the hazardous waste management industry do provide training for their employees, it is amongst the small operators and waste transport contractors where training needs are most urgently required. It is apparent that only a small proportion of workers receive external education and training for their work, either because appropriate courses are not readily available or workers are unable to attend them. Amongst the 12 Member States it is in Germany where great emphasis is being placed in the training of workers in the hazardous waste industry. For example, companies producing and treating hazardous waste must appoint a qualified 'agent-in-charge' who, along with numerous other duties, must monitor waste from production or delivery to disposal.

Recent accidents in the FRG have further emphasised the need for persons to be trained and made aware of the potential dangers during the collection, transport, delivery and marketing phases. Even where safety standards appear adequate, accidents have and will continue to occur because insufficient or inaccurate information was available concerning the nature of the waste load. Only if all waste loads are adequately monitored and accurately described and the personnel handling the waste are qualified and experienced in doing so can one give any assurances concerning the further improvement of public and environmental protection from hazardous wastes disposal.

5.3 Recommendations

Definition of Hazardous Waste

In order to achieve a greater degree of harmonisation over the management of hazardous waste it is recommended that the proposal for a Council Directive on hazardous waste is adapted. Furthermore, an industry or process based classification could be used to supplement the existing definition contained in the Toxic and Dangerous Waste Directive. It is noted that some Member States have already followed this route and thus, experience of applying this method is readily available to other Member States and can be used to develop a consensus over the nature of any process/industry class system.

The following recommendations concerning Health and Safety at Work and Environmental Legislation and Practice have been grouped according to specific target areas, these being legislation, (national and EC), regulatory agencies/plant operators and research community.

Legislation on Health and Safety at Work and Practice

Legislation

- In those Member States where the effectiveness of health and safety provisions are limited in scope either in relation to the definition of worker/workplace or in terms of the size of the establishment, amendments to the legislation should be made so as to ensure adequate protection for all workers.

- The forthcoming amendments to the national health and safety legislation of Member States relating to health surveillance of the workforce should encompass those individuals working with hazardous waste. Emphasis should be placed on workers in contact with complex mixtures of hazardous substances. The effectiveness of preventive measures should be monitored through medical surveillance and deficiencies to current health and safety provision, leading to demonstrably adverse health effects should be rectified.

- There is no urgent need to explore the role and nature of penalties in the enforcement of health and safety regulations.

Regulatory Agencies/Plant Operators

We have identified three primary areas :-

- Recent action by some Member States towards the use of preventative methods of reducing worker exposure to hazardous wastes for example, by the use of automated plant and remote handling procedures should be adopted by all Member States.

- The training and information obligations of employer and employees who have contact with hazardous wastes, under national regulations, should be more rigorously enforced at all levels. Emphasis should be given to issues such as the necessity and proper use of safety equipment and clothing when handling hazardous waste, and the need for safe procedures during the transportation of hazardous waste.

- Emergency services (e.g. fire services, doctors, police etc) should have specialist training if they are to attend accidents involving hazardous waste, so that they are aware of the most appropriate control measures and medical treatment requirements to mitigate the danger and damage. The actions of staff at hazardous-waste producing or treating facilities and the emergency services should be co-ordinated in order to maximise the resources and expertise available.

In order to bring about improvements of inspection and enforcement of legislation, the following actions are also desirable :-

- Recent trends in some Member States for worker participation in the formulation and implementation of safety policy should be encouraged.

- An assessment should be made of successful (company instigated) training programmes and ways of applying such programmes to hazardous waste handling facilities throughout Europe.

- Special arrangements, perhaps on a collective or association basis, should be made for small companies to assist them in making training and medical provision available to their employees.

- Information and/or technology exchange on health and safety issues and performance within the hazardous waste industry should be encouraged both to assist inexperienced Member States and to deter under-resourcing of the health and safety enforcement bodies by Member State governments.

Environmental Legislation and Practice

Legislation

- To improve enforcement and achieve consistent standards, some aspects of the legislation may need tightening up, i.e. effective compliance with certain aspects of the EC legislation needs to be defined (e.g. the nature and frequency of inspection of hazardous waste sites needs defining).

- If hazardous waste management is to be effectively planned, the competent bodies in some Member States need additional powers so that they have greater control over the movement of waste and how such waste should be disposed of, or treated, within their boundaries of responsibility.

- The penalties of illegal disposal need to be reviewed and increased to such a level where they discourage illegal disposal, for example in Spain, Italy and Greece, and are commensurate with the degree of environmental damage caused. Where contractors repeatedly 'infringe' waste regulations or 'codes of practice' the authorities should have powers to prohibit their business operation.

The following improvements on current practice are put forward for further consideration :-

- The term 'storage' should be precisely defined in national law as lack of information on the quantity of waste stored in some Member States is clearly increasing the uncertainty over waste arisings in general. For specific wastes, conditions for safe storage should also be defined.

- There is a need for an increased level of sampling and monitoring of waste arisings and particularly of disposal at co-disposal sites. It is evident, however, that more stringent monitoring of hazardous waste sites does not prohibit illegal disposal. In this respect some Member States would undoubtedly benefit if industry were required to submit a six-monthly or yearly 'disposal programme' to their competent body. This would detail the companies' arisings, material throughput, intended disposal routes and any planned process and production changes. With this information the competent body would be able to verify compliance with the disposal programme through 'spot-check' inspections of waste loads. It is also evident that some competent bodies would benefit if they had guidance on the initial setting of site conditions.

- Where problems of abandoned waste/chemical materials arise, it is recommended that, where not currently provided, the regional competent authority should have access rights to disused or recently vacated premises so as to ensure that appropriate measures can be taken before an accident or pollution incident occurs. In this respect, it may be necessary to set up regional funds which will cover the costs of clean up.

Regulatory Agencies/Plant Operators

- The resource allocation for enforcement purposes needs to be examined to identify where competent bodies are understaffed, poorly equipped and/or poorly trained and where they are simply operating inefficiently.

- The licence/authorisation conditions at many hazardous waste facilities need to be reviewed particularly with regard to waste acceptance, sampling and disposal procedures. Sites which have never been assessed for their suitability for receiving hazardous wastes should be considered as priority cases.

- Under Article 12 of the Toxic and Dangerous Waste Directive 78/319/EEC competent bodies are required to draw up plans for the disposal of toxic and dangerous waste. In the light of the poor level of compliance to date and the coverage of many of these plans, it is recommended that future waste disposal plans should be comprehensive, specifying waste arisings, disposal routes and disposal policy objectives, programmes and enforcement procedures. Such plans should also specify detailed policies and plans for the recycling and prevention of waste.

Additional points worthy of further consideration :-

- In some instances, bad practice and the transport (and transfrontier shipment) of waste is encouraged by large price differentials of disposal and a lack of appropriate infrastructure in some Member States. For example, in the UK contractors may decide to transport their waste considerable distances on public roads in order to benefit from cheap disposal in another area. Price controls and the setting up of regional treatment centres could be a useful mechanism to eliminate this practice particularly where the competent bodies are unable to control the movement of waste.

- Greater efforts should be made at national level for the organisation and operation of waste recylcing schemes and the eventual elimination of certain substances for manufacturing processes.

- Data collection sytems in Member States should be improved as current systems are largely inaccurate, cumbersome and rarely provide up-to-date information. In this respect, statistical analysis, monitoring and sampling of wastes should be harmonised throughout the EC.

- Where facilities handling/treating hazardous waste are recognised as examples of good practice, such facilities should be publicised through information exchange, visits, seminars etc to waste management authorities and companies throughout the EC. Two such examples of good practice are summarised in Annex I.

- Initiatives regarding new treatment capacity should be implemented with the co-operation of the waste producing industries and public authorities with the aim of producing centralised and not dispersed hazardous waste management systems.

Research Community

- The Toxic and Dangerous Waste Directive 78/319/EEC requires that Member States take steps to encourage, as a matter of priority, the prevention of toxic and dangerous waste, through recycling and material recovery. In this respect, greater efforts than hitherto, should be made at the national level for the organisation and operation of waste recycling schemes, the promotion of clean technologies and the eventual elimination of specific substances from manufacturing process and waste streams. It is recognised that the impact of such measures on hazardous waste management are in the long-term.

- The current provision for worker training, at all levels, needs to be reviewed and, where necessary, training courses should be developed and made available to all workers in the industry.

- Research into the development of new and less environmentally intrusive treatment and disposal methods should continue to be actively encouraged as not all wastes are currently treatable.

ANNEX I : CASE STUDY EXAMPLES

Landfilling of Hazardous Wastes

The requirement for extensive clean-up work on old refuse tips in many Member States, which were often designed and operated without any consideration of pollution of groundwater has, since the late 1970s, resulted in the implementation of a range of effective protective measures at landfill sites. These involve a detailed investigation of the site, sealing of the subsoil groundwater quality monitoring and strict operating procedures that ensure safe practice. The landfill case study described below is an example of good practice for the disposal of hazardous waste.

The landfill is designed to accept only pre-treated solid and dewatered wastes. In sealing the landfill, the prime purpose is to protect the groundwater against polluting leachate. Seepage water is captured by base filters and purified at an effluent treatment plant. Only when this has been done is it discharged into the main sewer outfall. For long-term protection natural sealing (clay) is preferred to artificial sealing, (use of PVC liners) the view being that a landfill for hazardous wastes cannot be made secure by artificial sealing alone. In this example the clay sealing is 4m thick. The sides of the site, above the clay, consist of gravel and sand. They are sealed by means of 0.60 m thick bentonite cement curtain walls to prevent inflows of groundwater and outflows of seepage water. These sub-level curtain walls, which form the landfill's boundary, are monitored during construction. During operation landfill they are constantly monitored during operation from observation wells inside and outside the landfill boundary.

A series of cells have been constructed within the landfill to accept different types of wastes. On arrival at the site all wastes are given a very careful preliminary check. On site materials management has to be decisive, unambiguous and cater for specific properties of the wastes. This is done by assessment, identification and allocation according to the management plan for individual types of waste. The options for the acceptance and analytical assessment of wastes are as follows :

a) visual assessment, odour test and rapid analysis (5 to 10 minutes)

b) operational or chemical analysis

Visual assessment, and subsequent space allocation is carried out by qualified and experienced personnel. Odour tests are intended mainly for liquid wastes. The purpose of rapid analysis is to assess wastes for the qualitative proportions of certain substances (chromate, cyanide, etc) and thus to decide where, in general terms, they should be deposited. Such an analysis can be backed up by "operational" analysis ; apart from allowing qualitative statements, this also permits a quantitative identification of the various admixtures.

The site has a planned capacity of some 3 million m³ and with an expected intake of between 120,000 and 150,000 tonnes the landfill will be operational for 18-20 years. This will be followed by recultivation of the site.

For further details on this case study see the National Report for Germany.

Integrated Hazardous Waste Treatment and Disposal Plant

This integrated hazardous waste treatment and disposal plant is located close to a residential area and therefore careful measures have been taken to ensure public and environmental protection from the initial planning stage through to commissioning and operation of the plant. During the course of its operation technologies and procedures have been continuously refined and adjusted to improve the level of safety and minimise the occurrence of accidents. The plant is capable of treating almost all types of chemical waste arising in industry. Certain types of waste are not treated, these are :

- Explosives
- Self-igniting substances
- Radioactive substances
- Hospital wastes (infectious wastes and pathological wastes)

The plant has the following capacity :

Incineration Plants

	Plant I	**Plant III**
	tonnes/operating hr	tonnes/operating hr
Pumpable wastes	2.5-3.0	3.0-3.5
Sludge	0.5-1.0	0.5-1.0
Solid wastes	0.5-1.0	1.0-1.5
Wastewater	1.5-2.0	1.5-2.0
Total wastes	4	5
Total wastewater	2	2

Oil Plant

Capacity approx 40000 tons/year with normal one-shift operation.

Inorganic Plant

Capacity approx 15000 tons/year with normal one-shift operation.

Component Plants

Tank-wagon emptying
Drum emptying
Oil-treatment plant
Inorganic treatment plant
Incineration plant I
Tank basin I for solvents
Storage hall for chemical wastes
Waste-water plant
Service building
Administration building with laboratory
Storage hall II and tank basin II
Soda lye tank, handling of toxic wastes and dosing plant
Drainage of rain and cooling water
Personnel building
Waste-oils plant
Incineration plant III
Reservoir for hydrochloric acid and soda lye
Drum-emptying plant II
Workshop
KaSa building
Extension of tank-wagon emptying
Turbine plant

These components have all been separately approved by the appropriate national agency responsible for environmental protection.

Operating conditions for the plant

Operating conditions and emission limits for the individual sections of the plant are specified. The important categories of emissions from the plant are :

a) Air emission from incinerators

b) Emission of solids (slags and filter dust) for dumping

c) Surface water

d) Drainage water and filter cake from inorganic plant

The air emissions limits for incinerator III are (under the authorisation granted in 1980):

Dust	< 100 mg/nm^3	
SO_2	< 750 mg/nm^3	
NO_X	< 300 mg/nm^3	
HCl	< 300 mg/nm^3	
HF	< 5 mg/nm^3	Emission limits are
Other halogenous hydrogens	< 5 mg/nm^3	1 hour average values and
Halogens (Cl_2)	< 25 mg/nm^3	must not be exceeded
Organic matter	< 300 mg/nm^3	
CO	< 150 mg/nm^3	
Odour equivalents	< 1000 $units/m^3$	

Air emissions for Incinerator I are :

Dust	< 150 mg/nm^3	
SO_2	< 1300 mg/nm^3	
NO_X	< 300 ppm	Emission limits are
HCl	< 600 mg/nm^3	1 hour average values and
HF	< 5 mg/nm^3	must not be exceeded
Halogens (Cl_2)	< 250 mg/nm^3	
Organic matter	< 300 mg/nm^3	
CO	< 1%	

For both Incinerator I and Incinerator III normal operating values are considerably lower than the limits applicable; for example, the example of a test result for plant III:

Dust	< 8 mg/nm³
SO_2	< 153 mg/nm³
NO_X	< 122 mg/nm³
HCl	< 87 mg/nm³
HF	0.4 mg/nm³
Halogens (Cl_2)	< 0.6 mg/nm³
CO	< 106 ppm (120 mg/nm³)

As new plant are introduced they may be subject to more stringent emission limits. A new authorisation, which will apply to the fourth incinerator, will reduce the limits for gaseous emissions for a number of parameters such as dust, HCl, HF, organic matter, etc.

Slags, filter dust and reaction residues from flue-gas scrubbing amount to about 20%, by weight of the quantity fed into rotating furnaces. Wastes of all three categories are deposited at the plants own controlled landfill site. The landfill authorisation was subject to specifically defined conditions.

The conditions include the comprehensive analysis of :

Percolate from slag section
Percolate from filter-dust section
Groundwater

Apart from the residual substances from the incineration plants referred to above, filter cake from the inorganic plant is also deposited. This is deposited in a special section, which is permanently covered with a strong plastic sheet to reduce the quantity of percolate from this section. This percolate is analysed in the same way as for the above.

Surface run-off

Run-off (rainwater) from hard-surfaced areas and roof areas is channelled to a common collection system which, via an oil separator and a delaying basin, is discharged in the sea. In association with the oil separator, samples are taken continuously and analysed for :

	Max Limit	
pH	8.5	
Conductivity	15	mS/cm
Temperature	35	C
Turbidity	400	NTV
TOC (total organic carbon)	90	ppm

If the maximum limit value is exceeded, the outlet from the delaying basins is closed and the discharge flow is reversed into a basin without outlet.

Process waste water

The inorganic plant produces a flow of process water originating from the dewatering process for metal hydroxide sludge (filter press). Following pH adjustment via a 24 hour delaying and post-precipitation basin, the water is discharged to the municipal purification plant.

The discharge requirements are specified for metals and organic and inorganic compounds.

	Max Limit		
pH	6.8 - 8.5		
Chromium (Cr) total	0.2	mg/l	Total metal
Lead (Pb)	-		max
Cadmium (Cd)	0.05	mg/l	5 mg/l
Zinc (Zn)	-		
Copper (Cu)	-		
Silver (Ag)	0.1	mg/l	
Nickel (Ni)	-		
Mercury (Hg)	3	ug/l	
Phenol	1	mg/l	
Chemical oxygen demand (COD)	1000	mg/l	
Cyanide, total (CN^-)	0.5	mg/l	
(BOD_5)	100	mg/l	
Mineral oil	10	mg/l	
Phosphate (as P)	10	mg/l	
Ammonia (as N)	25	mg/l	
Arsenic (As)	4	mg/l	
Fluorine (F^-)	10	mg/l	
Sulphate (SO_4)	5000	mg/l	

The inorganic plant produces a filter cake with a solids content of about 40%.

Supervision and approval of the plant is the responsibility of the National Agency of Environmental Protection. They have appointed an engineer, whose main duty is to supervise the plant.

Implementation of Supervision

Supervision is carried out both by notified inspections and by spot checks since under the appropriate law the supervisor may visit all parts of the enterprise, including its landfill site and talk to employees.

Inspections take the form of random investigations of physical conditions, including cleaning and general orderliness, and examination of files and print-outs from recording instruments, etc arising in connection with the enterprise's own monitoring activities.

In addition to the information collected from inspections on site, the plant operator forwards files, analysis data, etc to the Environmental Protection Agency. Comprehensive investigations of emissions from incineration plants are carried out twice a year. This material is also examined by the supervisor.

In addition to the above, the Environmental Protection Agency checks the site's own monitoring procedures.

If, on the basis of the inspections or examination of the material forwarded, evidence is found of a contravention of the requirements of environmental approvals or of other circumstances contrary to environmental legislation, an order is given for the matter to be rectified. Operation of the plant can thus be halted until a given situation has been rectified. In addition, information may be proffered to the police with a view to action before the courts.

Procedure in the event of accident, including on-call arrangements

If an accident or a breakdown in operation occurs at the plant causing contamination or involving a risk thereof, the plant operators (under the Environmental Protection Act) must immediately inform the supervisory authority. During normal working hours this can be done by approaching the Environmental Protection Agency's supervisor.

The Environmental Protection Agency has established an on-call system to deal with accidents or operational breakdowns outside working hours.

On-call duty is normally provided by the supervisor in co-operation with three persons from the local Environmental and Foodstuff Monitoring Unit, one of these four being on call day and night for a week at a time.

The person who is on call is equipped with a mobile telephone, so that he can always be telephoned and summoned.

In the event of a major accident outside normal working hours, a telephone list is also available so that the head office of the Environmental Protection Agency can be notified.

Whoever is called to an accident writes a report and the plant operator will if necessary be asked to submit a report on the accident and, where relevant, to take steps to avoid a similar occurrence.

Reporting

The supervisor submits a report to the Director of the Environmental Protection Agency each month on the supervision of the plant.

The report contains information on complaints received and the way in which they have been dealt with. It also describes important matters concerning protests, orders, bans, etc.

For further details on this case study see the National Report for Denmark.

ANNEX II : PARTICIPANTS OF THE COORDINATION MEETINGS

EXPERTS :

Ir H.S. BUIJTENHEK, G J KREMERS
TAUW Infra Consult B.V.
Handelskade 11
Postbus 479
7400 AL Deventer
Tel.: (05700) 99911

Mr Duncan BARDSLEY
ECOTEC Research and Consulting Ltd
Priory House
18 Steelhouse Lane
Birmingham B4 6BJ
Tel.: (021) 2369991

Dr G.U. FORTUNATI
Via V. Monti, 29
20123 Milan
Tel.: (031) 930861

Mr René GOUBIER
Agence Nationale pour la Récupération et L'Elimination des Déchets (ANRED)
2, Square Lafayette
B.P. 406
49004 Angers Cédex
Tel.: (41) 872924

Mr Jean-Pierre HANNEQUART
rue Baudoux 1,
1490 Beaurieux (Court-St-Etienne)
Tel.: (0932) 19588327

Ms Marie HANNEQUART
Rue Xhovemont, 20
4000 Liege

Mr Manuel DE LA VEGA DE LA ROSA, Mr Maximiliano JUNQUERA
TREISA s.a.
c/Condesa de Venadito, 1
28027 Madrid
Tel.: (191) 4031000

Mr Ulrich H. KINNER, Dr Manfred NICLAUSS
ECOSYSTEM GmbH
Kappellenstrasse 57
4005 Meerbusch 2
Tel.: (2159) 4050

EXPERTS (contd.) :

Mr Matthew **LYNCH**
EOLAS
Dublin Industrial Estate, 11, Unit 97
Ballymun Road
Dublin 9
Tel.: (01) 370101

Mr Mogens **PALMARK**
Chemcontrol A/S
Dagmarhus
1553 Copenhagen V
Tel.: (01) 141490

Mr José António Martins **REIS**
TECHINVEST
Rua Sanches Coelho, 3-6e
1600 Lisbon
Tel.: (01) 733001

RAPPORTEUR & ADVISOR :

Dr Richard C. **HAINES**
ECOTEC Research and Consulting Ltd
Priory House
18 Steelhouse Lane
Birmingham B4 6BJ
Tel.: (021) 2369991

REPRESENTATIVE OF THE FOUNDATION'S COMMITTEE OF EXPERTS :

Mr John **COFFEY**
Assistant Chief Engineering Advisor
Department of the Environment
Environmental Services Section
Custom House
Dublin 1
Tel.: (01) 728629

REPRESENTATIVE OF THE UNION GROUP :

Dr Werner **SCHNEIDER**
DGB-Bundesvorstand
Abteilung Umweltpolitik
Hans-Böckler-Strasse 39
4000 Düsseldorf 30
Tel.: 211 43011

REPRESENTATIVES OF THE EMPLOYERS' GROUP :

Mr Luc **DEURINCK**
CEFIC
Avenue Louise, 250
Bte. 71
B-1050 Brussels
Tel.: (02) 6402095

Mr JOURDAN
CEFIC
Avenue Louise
Bte. 71
B-1050 Brussels
Tel.: (02) 6402095

REPRESENTATIVES OF THE COMMISSION OF THE EC :

Mr J.-M JUNGER
DG XI
Commission of the European Communities
200, rue de la Loi
B-1049 Brussels
Tel.: (02) 2355442

Mr B. LEFEVRE
DG XI
Commission of the European Communities
200, rue de la Loi
B-1049 Brussels
Tel.: (02) 2355442

REPRESENTATIVES OF THE EUROPEAN FOUNDATION :

Mr Jorn PEDERSEN
Ms Hanna HANSEN
Ms Ann McDONALD
Solveig V. PETERSEN

ANNEX III : GUIDELINES FOR THE COMPLETION OF THE NATIONAL REPORTS

The aims of the study are :

- to provide the Foundation, the Community institutions and the Member States with a comprehensive picture and a comparative analysis of the state of the art and plans relating to the legislation, regulations and other measures in the Member States regarding safety issues in connection with the handling and monitoring of hazardous wastes, as part of the Foundation's overall effort towards safe waste management policies and practices which take into account the public concern in this area;

- to indicate possible alternatives to and improvements of the existing measures and practices in this area, e.g. by pointing to procedures which have been implemented successfully in some Member States;

- to provide the Community institutions with information which may assist them in their discussions on and definitions of future waste management policies.

CONTENT OF THE STUDY

As there is no agreed definition within the Community of the terms "hazardous wastes", "toxic and dangerous wastes" or "special wastes" the national definitions of such wastes will apply, and this is covered by the term "hazardous wastes" used in this document.

The findings of each national study will be presented in a report which will be divided into the following chapters and will include the information mentioned below under each of the chapter headings.

1. **Introduction**

This chapter will be very short (5-10 pages) and will focus on general background information. It will include :

- the national definition of hazardous wastes, its background and the problems which it may give rise to, e.g. in terms of interpretation.

- the national annual arisings of hazardous wastes as defined nationally or regionally, their sources and their disposal/treatment or re-use or re-cycling locally, regionally, nationally or abroad (export);

- the quality of hazardous wastes transported by road, rail, sea or internal waterways;

- the collection of data on hazardous wastes, the methodologies applied, their accuracy, their differences and the bodies in charge of the data collection.

2. Work-Safety Legislation and Regulations applying to Personnel involved, whether directly or indirectly, in the Handling and Monitoring of Hazardous Wastes

This chapter will describe and analyse the work safety legislation and regulations, their background and their enforcement, particularly in relation to :

- personnel employed at waste collection centres, waste disposal sites, waste treatment plants or plants generating or handling considerable amounts of hazardous wastes (e.g. for re-use or re-cycling);

- personnel involved in excavating and cleaning up all categories of contaminated land;

- drivers and other personnel involved in the transportation of hazardous wastes by road, rail, sea or internal waterways (including loading and unloading of wastes).

Where such work-safety legislation differs among regions, a description of the major regional differences, applying to each region, will be given, together with an explanation of the reasons for the differences. Equally, when there is scope for local interpretation and implementation of the legislation, an explanation will be given, relating, in particular, to whether this actually takes place and the differences to which such a system leads.

When describing the legislation its fields of application will be specified, e.g. whether it applies to dangerous goods, substances and wastes, to all wastes, irrespective of their nature, or only to specific categories of wastes. If there are areas which are not covered by the work-safety legislation on the handling of hazardous wastes but by other regulations, as it may be the case, for instance, with regard to wastes from research laboratories, workshops and hospitals, this situation will also be illustrated. Moreover, if the work-safety legislation applying to the handling of hazardous wastes is different from that applying to dangerous goods and substances a comparison will be made of the two sets of rules. Likewise, insofar as there are differences in the work-safety legislation applying to personnel working at waste-collection centres, waste disposal sites, contaminated land being excavated, wastes treatment plants etc, such differences will be highlighted and explained, and so will any major deviations (both nationally and regionally) from the general legislation on safety at work.

With a view to obtaining a picture of recent developments and trends regarding the work-safety legislation, each national report will indicate major changes introduced during the last 10-15 years, although focusing on the most recent ones, as well as those planned, and it will explain the reasons for those changes.

Insofar as other parts of the legislation (e.g. transport, environmental or industrial legislation) have an impact on the work-safety legislation in this area an information is available this impact will be explained and it will be established whether there is a need for harmonisation of rules and regulations.

This might be relevant, for instance, in relation to the transportation of hazardous wastes, be it within or between regions or between Member States. A conflict between work safety legislation of two different regions or Member States might also occur in this area.

Moreover, each national report will contain information on the implementation and enforcement procedures and the bodies/agencies in charge of implementing and/or enforcing the legislation and their respective responsibilities and roles in this process.

The final part of the chapter will focus on the assessment of the existing legislation in this area and the planned changes. An attempt will be made to highlight the problems encountered in relation to the legislation and its implementation, the possible shortcomings and enforcement difficulties and, where needed, possible improvements will be suggested. An assessment will also be made as to whether the existing legislation has been influenced by other considerations than the potential risks of hazardous wastes, and whether it needs to be modified for this reason and to be made more specific.

3. National Safety Legislation and Regulations applying to Waste Disposal Sites, Treatment Plants and other Establishments handling Hazardous Wastes.

This chapter will describe and analyse the above-mentioned legislation and regulations, their background and enforcement.

If such a legislation or regulations differ on major points from one region to the other, a description of such regional differences or of the whole set of regulations applying to each region will be given, together with an explanation of the reasons for the differences. Likewise, when there is scope for local interpretation and implementation of the legislation, and explanation will be given, relating to the extent to which this actually takes place and to what it leads.

The great majority of the Member States have carried through a framework legislation on control and disposal of hazardous wastes, either prior to or in implementing Directive 78/319/EEC on Disposal of Toxic and Dangerous Waste. The essential objectives of this directive is the protection of human health and the safeguarding of the environment against harmful effects caused by the collection of toxic and dangerous waste, as well as its transportation, treatment, storage and disposal. This principle objective is reflected in the legal provisions of the Member States, but there is nevertheless considerable divergence between the content of the national legislations and their field of application and consequently between the implementation, control and enforcement. Some of the main differences are linked to the policy objectives persued, the definition of hazardous wastes, the focus of control (whether it is on waste streams or on handling and disposal facilities) and the application of control directly to waste producers. The fact that the local authorities almost everywhere in the Community have the major responsibility for implementation and enforcement of the legislation also tends to result in uneven application and varying standards of control as well as other problems.

These particularities and differences, together with the underlying factors, will be explained in each national report in connection with the description and analysis of the existing legislation, its implementation and enforcement. In this context the reports will aim to obtain a picture of recent developments and trends regarding the safety legislation in this area. Each report will therefore indicate major changes introduced during the last 10-15 years, with emphasis on the most recent ones, as well as those planned, and explain the reasons for these changes.

If other relevant safety provisions have been introduced in addition to those included in the above-mentioned framework legislation and implementing regulations such provisions will, of course, also be included in the description and analysis of the existing legislation in this area, so as to obtain a full picture of the situation. The same applies if there are special provisions regarding the disposal, storage and treatment of hazardous wastes falling outside the scope of Directive 78/319/EEC, e.g. hospital waste and other toxic and dangerous wastes covered by specific Community rules.

Finally, if other parts of the legislation (e.g. health and industrial legislation) have an impact on the safety regulations in this area this impact will be explained, and an attempt will be made to establish whether there is a need for harmonisation of rules and regulations.

When describing and analysing the legislation in this field, its implementation and enforcement, each report will follow, although in a slightly modified form, the outline of Directive 78/319/EEC, i.e.:

(i) the scope of the legislation and approaches to control in the Member States;

(ii) identification/definition of hazardous wastes

(iii) ensuring controlled disposal of hazardous wastes, separation of such wastes from other matter or residues, packaging and labelling requirements, records and identification of such wastes in respect of each sites;

(iv) allocation of responsibilities to competent authorities in relation to planning, organisation, authorisation and supervision;

(v) authorisation of disposal facilities;
- provisions
- progress in authorising undertakings;
- controlling and monitoring local authority authorisation;

(vi) authorisation or control measures relating to the transportation of hazardous wastes;

national and regional levels, and not only between different types of establishments, but also between establishments within the same category. This is essentially due to the fact that in most cases it is the local authorities which have the major responsibility for authorising and supervising the operations of facilities for storage, treatment and disposal of hazardous wastes.

For obvious reasons, each national report will have to focus on a limited number of such regulations, i.e. apply a case-study orientated approach, and a selection is therefore needed. This selection must necessarily be representative in the national/region context so as to ensure reliable comparisons. Hence, a detailed discussion and a decision on the criteria to be applied to the selection will be required at the first co-ordination meeting on this project.

A representative sample of facilities/establishments and internal safety regulations, however, will only illustrate the current practices and will therefore be supplemented by a few facilities/establishments, which are considered the most advanced in their field in relation to safety measures, so as to also indicate the scope for possible improvements.

As part of their description and analysis of the internal safety regulations the reports will aim to answer the following questions :

- Who issued the licence? What were the specific conditions in the licence in relation to safety measures? How were these conditions implemented into the safety regulation, and who prepared this regulation?

- Was the safety regulation approved by a public authority, and if so, by a government body, a regional or a local authority?

- Is this facility/establishment subject to supervision and inspection from an outside body? If so, which authority is in charge of this supervision and inspection, and how is it organised? Who carries out the inspection? How often does it take place? Has it ever led to any major changes in the safety measures, and if so, which changes?

- How is the internal supervision of safety organised? Are representatives of the personnel involved? Has it given rise to any problems in recent years?

(vii) monitoring and supervision of disposal facilities and undertakings engaged in transportation of hazardous wastes;

(viii) planning for the disposal of hazardous wastes;

(ix) emergency measures.

This outline may have to be further modified in the light of other relevant safety provisions not included in the above-mentioned framework legislation and implementing regulations. There may also be a need for a separate item dealing with risk assessment, both in relation to hazardous wastes as such to disposal facilities. A decision in this respect will be taken following the discussion at the first co-ordination meeting on the study.

The final part of the chapter will focus on the assessment of the existing legislation and the planned changes. The problems encountered in relation to the various provisions, their implementation and enforcement will be highlighted, together with possible shortcomings and loopholes and possible improvements. In the light of the above-mentioned description and analysis of the legislation and its implementation, this part will also include a consideration of the role of planning regulations and site licensing in controlling not only the location/siting of waste disposal facilities, but also the degree of control over what wastes are accepted, treatment carried out and disposal method used on site. Moreover, if the legislation in this area differs on major points from the safety provisions relating to dangerous goods and substances, a comparison will be made between the two sets of rules.

4. Internal Safety Regulations at Waste Disposal Sites, Treatment Plants and other Establishments handling Hazardous Wastes (e.g. collection centres and certain categories of industrial plants)

This chapter will focus on a description and analysis of internal safety regulations at waste disposal facilities and other establishments handling hazardous wastes.

It must be assumed that the primary objective of these regulations is to fulfil the specific requirements in the legislation referred to under points 2 and 3 above, and, if so, they will reflect the actual implementation of parts of this legislation. Nevertheless, they are likely to differ considerably, both at the

- Does the safety regulation imply extensive training of the personnel, and if so, how is the training organised? What is the educational background of the personnel?

- Do foreign nationals take part in the training programmes, and do they have access to written safety instructions in their mother tongue and/or does there exist other means of communication which ensure that they are always informed of the safety instructions?

- Are representatives if the personnel involved in the formulation and monitoring of the provisions in the safety regulation, and if so, how is their involvement organised, e.g. as members of a a consultative body?

- Which safety measures are the most difficult to comply with for the facility/establishment, and why? Which safety measures are the most difficult to comply with for the personnel, and why? Are there areas of safety where it is difficult to obtain the cooperation of the personnel, and why?

- Which first aid facilities exist at the facility/establishment? Does it have contact with outside emergency facilities (e.g. hospitals, fire brigade, police) and , if so, do they know about the nature of the wastes treated or disposal and the measures needed in case of accidents?

- Has the internal safety regulations been changed in the last five years, and why? Have accidents at the facility/establishment resulted in changes of the regulations?

- Are there any plans for changing the regulation and, if so, which changes are likely to be introduced?

The analysis of the internal safety regulations and the answers to the above-mentioned questions should provide a good basis for comparisons between the measures and practices at various facilities/establishments as well as a reasonable picture of the implementation of the national legislation and the difficulties encountered in this report. An assessment will be made of the current practices and, if needed, the scope for improvements, e.g. by pointing to example of goods practices at other facilities and establishments. Moreover, an attempt will be made to explain the reasons for difficulties in implementing and enforcing the legislation, e.g. structural, economic, shortcomings or loopholes in the legal provisions, and how these problems may be overcome, thereby leading to better practices.

Information on the Nature of the Wastes, the Potential Risks and the Measures Required in Case of Accidents

This chapter will concentrate on a description and analysis of the practices followed, particularly at waste disposal facilities, in relation to preliminary and control analyses before and at the arrival of wastes, together with the technical and personnel resources available for this purpose. Information on the criteria for acceptance of different types of waste and whether they are based on technical norms (e.g. corrosion), environmental considerations (e.g. emission standards) or norms relating to work safety and public health, as well as the criteria for the selection of waste disposal facilities will also be included. Furthermore, waste notification systems will be analysed, as well as the ability to identify and assess potential risks from, for example, the mixing of various waste streams. Monitoring requirements and procedures will also be examined.

Substantial differences between the legal provisions in each Member State, their implementation and enforcement are certain to prevail in this area, leading to a diversity of procedures and practices. Even within one Member State, the procedures and practices will vary considerably, insofar as the major responsibility for issuing licences and exercising control is left to the local authorities. For this reason, a case-study orientated approach will be adopted, similar to the one mentioned under point 4 above. It will even be possible to use the same sample, if a criteria selection is made, thus enduring that it will be sufficiently representative in its illustration of typical and advanced practices in each national and regional context.

Part of this report will focus on a assessment of the possible ways and means of introducing a Community system, providing all persons involved in the handling and monitoring of hazardous wastes with sufficient and clear information on these wastes and on the measures required in case of accidents, e.g. by use of waste cards accompanying the wastes from their generation to their final disposal.

6. Summary

In this final chapter the reports will summarise the analysis of the information referred to under points 2 - 5 above and, in particular, the findings and conclusions. This chapter will not exceed ten pages, and will be drafted in such a way that it can be read independently of the other chapters.

THE EUROPEAN FOUNDATION FOR THE IMPROVEMENT OF LIVING AND WORKING CONDITIONS

SAFETY ASPECTS RELATING TO THE HANDLING AND MONITORING OF HAZARDOUS WASTES

Luxembourg: Office for Official Publications of the European Communities

1989 - p.- 160 x 235 mm

ISBN: 92-826-0140-4

Catalogue Number: SY-57-89-330-EN-C

Price (excluding VAT) in Luxembourg :
ECU 12.50

F/89/26/EN

rice (excluding VAT) in Luxembourg:
CU 12.50